Civil Engineering Materials
土木工程材料

董 侨 陈雪琴 主 编
高 英 胡建英 陈先华 徐光霁 副主编

东南大学出版社
SOUTHEAST UNIVERSITY PRESS
·南京·

内容简介

This book is to introduce the basic knowledge, tests, and designs of civil engineering materials, covering the fundamental mechanical and physical properties of materials; properties, tests, and gradation designs of aggregates; production, composition, hydration, properties, and tests of lime and cement; composition, tests, and design of cement concrete; mechanisms, properties and design of inorganic binder stabilized material; properties, tests, and grading of asphalt; composition, properties, tests and designs of asphalt mixture; properties, treatments, tests, and selections of steel. The specifications in China and several key specifications from the US are introduced. This book can be used as a textbook or a reference book for undergraduate students, graduate students, and professionals in the field of civil, pavement, bridge, geotechnical, and environmental engineering.

图书在版编目(CIP)数据

土木工程材料 = Civil Engineering Materials：英文 / 董侨,陈雪琴主编. — 南京：东南大学出版社，2022.8
ISBN 978-7-5766-0226-5

Ⅰ.①土… Ⅱ.①董… ②陈… Ⅲ.①土木工程-建筑材料-英文 Ⅳ.①TU5

中国版本图书馆 CIP 数据核字(2022)第 157518 号

责任编辑：宋华莉　　责任校对：张万莹　　封面设计：余武莉　　责任印制：周荣虎

Civil Engineering Materials
土木工程材料

主　　编	董　侨　陈雪琴
出版发行	东南大学出版社
社　　址	南京市四牌楼 2 号(邮编：210096　电话：025-83793330)
经　　销	全国各地新华书店
印　　刷	南京玉河印刷厂
开　　本	700mm×1000mm　1/16
印　　张	19.5
字　　数	376 千字
版　　次	2022 年 8 月第 1 版
印　　次	2022 年 8 月第 1 次印刷
书　　号	ISBN 978-7-5766-0226-5
定　　价	78.00 元

本社图书若有印装质量问题，请直接与营销部联系，电话：025-83791830。

Preface

Civil engineering materials are the fundamentals for architectural design, structural design, buildings, bridges, pavements, construction management, etc. The course of civil engineering materials is a fundamental course for civil engineering, pavement engineering, or other related majors. Civil engineering materials cover multiple disciplines including material science with a specific focus on concrete, metal, polymer, and composites, the mechanics such as theoretical, structural, fracture, and fatigue mechanics, and the knowledge of testing, mathematics, statistics, etc. The objectives of this book are to help students understand the definition, classifications, properties, and testing of civil engineering materials, the storage, transportation, and protection of materials, as well as the design of materials so that students can select and use civil engineering materials.

The course of civil engineering materials at Southeast University was established around a century ago. In 1919, Dr. Yisheng Mao obtained a Ph. D. degree from Carnegie Mellon University. He was the first Ph. D. of engineering at Carnegie. In 1923, he came back to China and established the Department of Civil Engineering at Southeast University. In the same year, Mr. Zhihong Lu graduated from Tokyo University with the highest GPA. He then joined Southeast University and established the course and lab for civil engineering materials in 1928. The content of this course has been updated with the development of technologies in civil engineering materials. This book aims to cover the most recent theories and practices in civil engineering materials.

Many test and design specifications of civil engineering materials in China were established by referring to the specifications from many countries in North America, Europe, and Asia. This book provides related test and design specifications in China. Some specifications of the American Society for Testing and Materials (ASTM) International, the American Association of State Highway and Transportation Officials (AASH-

TO), and other countries are also mentioned. This book can be used as a reference book for undergraduate students, international students, and professionals in the area of civil, pavement, bridge, and geotechnical engineering.

It is impossible to cover all the civil engineering materials in one single book. This book focuses on the fundamentals and principles of the knowledge related to civil engineering materials. Dr. Qiao Dong and Dr. Xueqin Chen drafted the whole book. Dr. Ying Gao, Dr. Guangji Xu, Dr. Xianhua Chen, and Dr. Jianying Hu helped to revise chapters five to eight. The main contents of each chapter are listed as follows:

- The first chapter introduces the basic properties of civil engineering materials, including the classifications, mechanical properties, density, water-related properties, durability, workability, and variability of test results.
- The second chapter introduces the classification, production, physical and mechanical properties of coarse and fine aggregates, gradation curves, and gradation designs.
- The third chapter introduces the production, slaking, hardening, and properties of lime; and the production, hydration, properties, corrosion, and supplementary materials of cement.
- The fourth chapter introduces the composition, workability, strength, deformation, durability, mix design, and admixtures of cement concrete.
- The fifth chapter introduces the classifications, properties, strength, shrinkage, and mix designs of inorganic binder stabilized materials.
- The sixth chapter introduces the production, composition, properties, aging, modification, penetration grade, viscosity grade, and performance grade of asphalt binder.
- The seventh chapter introduces the classification, structure, properties, volumetric parameters, mix design, and construction of asphalt mixtures.
- The eighth chapter introduces the production, mechanical properties, hot and cold treatment, types, and corrosion of steel.

Contents

1 Fundamentals of Civil Engineering Materials ········· 001
- 1.1 Background ········· 001
- 1.2 Classifications ········· 002
- 1.3 Mechanical Properties ········· 004
  - 1.3.1 Deformation ········· 005
  - 1.3.2 Stress and Strain ········· 007
  - 1.3.3 Strength ········· 008
  - 1.3.4 Elasticity ········· 008
  - 1.3.5 Plasticity ········· 009
  - 1.3.6 Toughness ········· 009
  - 1.3.7 Brittleness ········· 010
  - 1.3.8 Stiffness ········· 010
  - 1.3.9 Ductility ········· 010
  - 1.3.10 Hardness ········· 011
  - 1.3.11 Viscosity ········· 011
  - 1.3.12 Viscoelasticity ········· 012
- 1.4 Non-mechanical Properties ········· 013
  - 1.4.1 Density ········· 013
  - 1.4.2 Thermal Properties ········· 015
  - 1.4.3 Surface Properties ········· 018
  - 1.4.4 Water-related Properties ········· 019
  - 1.4.5 Durability ········· 021
  - 1.4.6 Workability ········· 022
- 1.5 Variability ········· 022
- Questions ········· 023

2 Aggregates ... 024

- 2.1 Production of Aggregates ... 024
  - 2.1.1 Aggregate Sources ... 024
  - 2.1.2 Types of Rocks ... 025
  - 2.1.3 Physical Properties of Rocks ... 027
  - 2.1.4 Mechanical Properties of Rocks ... 028
  - 2.1.5 Chemical Properties of Rocks ... 029
  - 2.1.6 Crusher ... 029
- 2.2 Physical Properties of Coarse Aggregates ... 030
  - 2.2.1 Sampling ... 030
  - 2.2.2 Sieve Analysis and Gradation ... 031
  - 2.2.3 Aggregate Size ... 033
  - 2.2.4 Moisture Condition ... 034
  - 2.2.5 Density ... 035
  - 2.2.6 Unit Weight and Voids in Aggregates ... 038
  - 2.2.7 Angularity and Flakiness ... 040
  - 2.2.8 Fractured Faces and Crushed Particle ... 042
- 2.3 Mechanical Properties of Coarse Aggregates ... 042
  - 2.3.1 Crushing Value ... 043
  - 2.3.2 Impact Value ... 043
  - 2.3.3 Polished Stone Value ... 043
  - 2.3.4 Abrasion Value ... 045
  - 2.3.5 Soundness ... 045
  - 2.3.6 Alkali Reaction ... 046
- 2.4 Properties of Fine Aggregates ... 047
  - 2.4.1 Density ... 047
  - 2.4.2 Unit Weight and Void Ratio ... 049
  - 2.4.3 Angularity ... 050
  - 2.4.4 Fineness Modulus ... 050
  - 2.4.5 Sand Equivalency ... 051
  - 2.4.6 Methylene Blue Test ... 052
- 2.5 Filler ... 052

Contents

2.6　Gradation Design ········· 053
    2.6.1　Gradation Curves ········· 053
    2.6.2　Gradation Theory ········· 056
    2.6.3　Gradation Design ········· 057
Questions ········· 059

3　Inorganic Binding Materials ········· 062

3.1　Lime ········· 062
    3.1.1　Production ········· 062
    3.1.2　Slaking and Hardening ········· 064
    3.1.3　Properties ········· 065
3.2　Production of Cement ········· 067
    3.2.1　Classification ········· 067
    3.2.2　Production ········· 068
    3.2.3　Composition ········· 069
3.3　Hydration of Cement ········· 071
    3.3.1　Hydration Products ········· 071
    3.3.2　Hydration Process ········· 073
    3.3.3　Influencing Factors ········· 074
3.4　Properties of Cement ········· 077
    3.4.1　Density ········· 077
    3.4.2　Fineness ········· 078
    3.4.3　Consistency ········· 079
    3.4.4　Setting Time ········· 079
    3.4.5　Soundness ········· 080
    3.4.6　Strength ········· 080
    3.4.7　Hydration Heat ········· 082
    3.4.8　Voids ········· 083
3.5　Corrosion ········· 083
    3.5.1　Soft Water Corrosion ········· 084
    3.5.2　Sulfate Attack ········· 084
    3.5.3　Magnesium Corrosion ········· 085
    3.5.4　Carbonation ········· 085
    3.5.5　Acid Corrosion ········· 086

    3.5.6 Measures ········· 086
  3.6 Supplementary Materials and Blended Cement ········· 087
    3.6.1 Supplementary Materials ········· 087
    3.6.2 Blended Cement ········· 089
  Questions ········· 093

4 Cement Concrete ········· 094

  4.1 Classification and Composition ········· 095
    4.1.1 Classification ········· 095
    4.1.2 Composition ········· 095
  4.2 Workability of Fresh Concrete ········· 098
    4.2.1 Workability ········· 098
    4.2.2 Influencing Factors ········· 102
    4.2.3 Measures to Improve Workability ········· 104
  4.3 Properties of Hardened Concrete ········· 105
    4.3.1 Mesoscale Structure ········· 105
    4.3.2 Strength of Concrete ········· 106
    4.3.3 Strength Influencing Factors ········· 112
    4.3.4 Stress-strain Relationship ········· 116
    4.3.5 Shrinkage ········· 118
    4.3.6 Durability of Concrete ········· 120
  4.4 Admixtures ········· 123
    4.4.1 Water Reducers ········· 123
    4.4.2 Air Entrainer ········· 124
    4.4.3 Setting Adjuster ········· 124
    4.4.4 Shrinkage Reducer ········· 125
  4.5 Mix Design ········· 126
    4.5.1 Information Collection ········· 126
    4.5.2 Initial Formula ········· 126
    4.5.3 Basic Formula ········· 130
    4.5.4 Lab Formula ········· 131
    4.5.5 Field Formula ········· 132
    4.5.6 Water Reducer Adjustment ········· 133
  Questions ········· 133

Contents

5 Inorganic Binder Stabilized Materials ... 135

- 5.1 Applications and Classifications ... 135
  - 5.1.1 Applications ... 135
  - 5.1.2 Classifications ... 137
- 5.2 Modification Mechanism ... 139
  - 5.2.1 Cement Stabilized Material ... 139
  - 5.2.2 Lime Stabilized Material ... 142
  - 5.2.3 Lime and Industrial Waste Stabilized Material ... 144
- 5.3 Mechanical Properties ... 148
  - 5.3.1 Unconfined Compression Test ... 148
  - 5.3.2 Resilience Modulus Test ... 152
  - 5.3.3 Split Tension Test ... 152
  - 5.3.4 Flexural Strength Test ... 154
  - 5.3.5 Fatigue Performance ... 155
- 5.4 Shrinkage ... 157
  - 5.4.1 Tests of Shrinkage ... 158
  - 5.4.2 Drying Shrinkage Mechanism ... 159
  - 5.4.3 Thermal Shrinkage Mechanism ... 160
- 5.5 Mix Design ... 161
  - 5.5.1 Design Procedure ... 162
  - 5.5.2 Requirements of Materials ... 165
  - 5.5.3 Requirements of Compaction ... 172
- Questions ... 172

6 Asphalt ... 173

- 6.1 Classification and Production ... 173
  - 6.1.1 Classification ... 173
  - 6.1.2 Production ... 174
- 6.2 Composition and Structure ... 176
  - 6.2.1 Composition ... 176
  - 6.2.2 Colloidal Structure ... 177
- 6.3 Properties ... 179
  - 6.3.1 Physical Properties ... 179
  - 6.3.2 Penetration ... 179

6.3.3	Viscosity	180
6.3.4	Softening Point	182
6.3.5	Brittle Point	183
6.3.6	Ductility	183
6.3.7	Adhesion	184
6.3.8	Durability	184
6.3.9	Safety	185

6.4 Temperature Susceptibility ………………………………………………… 186
    6.4.1 Temperature Dependency ………………………………………… 186
    6.4.2 Time-temperature Equivalency …………………………………… 187
    6.4.3 Penetration Index …………………………………………………… 187

6.5 Aging and Modification ……………………………………………………… 189
    6.5.1 Aging …………………………………………………………………… 189
    6.5.2 Modification …………………………………………………………… 191

6.6 Penetration and Viscosity Grading ………………………………………… 192
    6.6.1 Penetration Grading ………………………………………………… 192
    6.6.2 Viscosity Grading …………………………………………………… 196

6.7 Performance Grading ………………………………………………………… 197
    6.7.1 Equipment …………………………………………………………… 197
    6.7.2 Testing Temperature ………………………………………………… 202
    6.7.3 PG Grades …………………………………………………………… 203
    6.7.4 PG Tests ……………………………………………………………… 204
    6.7.5 Selection of PG Grades …………………………………………… 205

Questions …………………………………………………………………………… 208

7 Asphalt Mixtures ……………………………………………………………… 209

7.1 Classification ………………………………………………………………… 209
    7.1.1 Type of Mixture ……………………………………………………… 209
    7.1.2 Temperature ………………………………………………………… 210
    7.1.3 Gradation …………………………………………………………… 211
    7.1.4 Aggregate Size ……………………………………………………… 212

7.2 Composition and Strength ………………………………………………… 213
    7.2.1 Composition ………………………………………………………… 214
    7.2.2 Strength Parameters ………………………………………………… 215

		7.2.3	Influencing Factors	217
	7.3	Properties and Tests		220
		7.3.1	High-Temperature Stability	220
		7.3.2	Low-Temperature Cracking	229
		7.3.3	Fatigue Performance	231
		7.3.4	Moisture Susceptibility	233
		7.3.5	Friction	235
	7.4	Volumetric Parameters		236
		7.4.1	Volumes in Mixture	236
		7.4.2	Calculation of Parameters	239
	7.5	Marshall Mix Design		244
		7.5.1	Background	244
		7.5.2	Procedures	245
	7.6	Superpave Mix Design		250
		7.6.1	Background	250
		7.6.2	Procedures	251
	7.7	Typical Mixtures		256
		7.7.1	Dense-Graded Mixtures	256
		7.7.2	Gap-Graded Mixtures	257
		7.7.3	Open-Graded Mixtures	258
		7.7.4	Poured Asphalt Mixtures	259
	7.8	Construction		260
		7.8.1	Manufacture	260
		7.8.2	Transportation	261
		7.8.3	Paving	261
		7.8.4	Compaction	262
		7.8.5	Recycling	262
	Questions			263
8	**Steel**			264
	8.1	Production		264
		8.1.1	Iron Production	265
		8.1.2	Steel Production	265

8.2	Classification		266
	8.2.1	Deoxygenation	266
	8.2.2	Chemical Composition	267
	8.2.3	Applications	268
	8.2.4	Steels in Civil Engineering	269
8.3	Chemical Composition		272
	8.3.1	Influence of Elements	272
	8.3.2	Metallography	275
8.4	Mechanical Properties		275
	8.4.1	Strength	276
	8.4.2	Plasticity	277
	8.4.3	Cold Bending Property	278
	8.4.4	Impact Toughness	279
	8.4.5	Hardness	279
8.5	Heat and Cold Treatment		280
	8.5.1	Heat Treatment	280
	8.5.2	Welding	282
	8.5.3	Cold Working	283
8.6	Typical Structural Steel		284
	8.6.1	Hot-Rolled Steel Bar	284
	8.6.2	Cold-Rolled Ribbed Bar	286
	8.6.3	Hot-Rolled Reinforcing Bar for Prestressed Concrete	286
	8.6.4	Cold-Drawn Low Carbon Steel Bar	287
	8.6.5	Prestressed Steel Wire, Indented Steel Bar and Steel Strand	288
	8.6.6	Steel Sections	289
8.7	Corrosion and Protection		292
	8.7.1	Corrosion	292
	8.7.2	Protection	292
Questions			293
Reference			294

1 Fundamentals of Civil Engineering Materials

1.1 Background

Civil engineering materials have shaped human history since the dawn of civilization. Many significant achievements in the production and use of civil engineering materials have been made since ancient times. Around 3000 years ago, ancient Chinese started using lime in building houses and the Great Wall which used 100 million cubic meters of masonry materials (Figure 1.1). The Luoyang Bridge in Quanzhou, Fujian Province, was built of stone more than 900 years ago, and each stone weighed more than 200 tons.

Figure 1.1 The Great Wall

Today, the technology of civil engineering materials is still developing. Figure 1.2 shows a cable-stayed bridge, which is one of the 70 bridges over the Yangtze River in China. The cable is 577 m long and the diameter of the cross-section is the same height as an adult. Inside the cable, each of the 7 mm diameter steel wires bears 7 tons. Since civil engineering consumes a great number of materials, using better quality and more economical materials is of great importance.

 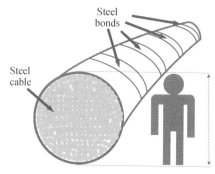

Figure 1.2 A cable-stayed bridge

1.2 Classifications

Civil engineering materials are the general terms for all types of materials used in civil engineering. The performance, costs, and the needed amount in a structure of different materials vary greatly. Therefore, the correct selection and design of materials are of great significance to the safety, durability, and costs of structures. A good civil engineer knows how to select and design the best materials for different structures, including buildings, highways, railways, bridges, tunnels, airports, etc., as shown in Figure 1.3. Only with a good understanding of the material properties can a civil engineer accurately determine the dimensions of structures and create advanced structural forms. The whole process of a civil engineering project is essentially the process of gradually turning civil engineering materials into buildings according to the design requirements. It involves the selection, transportation, storage, testing, designing, and processing of materials. The cost of materials accounts for about 50% to 60% of the total cost of a project. Usually, the performance of materials is maximized considering structure, durability, construction, and economic factors. Design, construction, and management are closely related. Fundamentally, materials are the foundation, and they determine the form of civil structures and construction methods. The emergence of new materials can lead to changes in the form of civil structures, improvements in design methods, and innovations in construction techniques.

Figure 1.3 Different civil engineering structures

The civil engineering materials mainly include aggregates, inorganic and organic binders, mixtures, steel, and wood. Aggregate materials include rocks from the upper layers of the earth's crust obtained by natural weathering such as natural gravel or artificial mining and crushing. These materials can be used directly in civil engineering structures or as the skeleton for cement concrete and asphalt mixtures. Inorganic binding materials mainly include lime and cement. Lime or cement mortar is an important binding material for masonry structures. Cement concrete is the main material for many large-volume structures including buildings, highways, dams, etc. Inorganic binder stabilized materials are used as the main material type of pavement base course. Organic binding materials are mainly asphalt binders which can be used as sealing or waterproofing materials. More importantly, they are mixed with aggregates to produce various asphalt mixtures for pavements. Steel is an important material for bridges, steel structures, and reinforced concrete or pre-stressed reinforced concrete structures. Wood is mainly used for house building, frameworks, and decorations in civil engineering.

Civil engineering materials can be classified as organic materials, inorganic materials, and composite materials according to their composition. Organic materials include asphalt, wood, and polymers; inorganic materials include metal, aggregates, and cement; composite materials generally include more than one type of material, such as reinforced concrete. According to the applications, civil engineering materials can be divided into two categories, namely the structural materials contributing to the capacity and safety of structures and functional materials contributing to the function and quality of the structures. It is noted that materials may have multiple functions. Structural materials mainly refer to the materials used for the load-bearing parts such as beams, slabs, columns, foundations, frames, walls, arches, pavement courses. The main performance requirements of these materials are strength and stiffness. The structural materials mainly include brick, stone, cement concrete, steel, reinforced concrete, prestressed reinforced concrete, asphalt concrete, and inorganic binder stabilized materials. Functional materials are mainly non-load-bearing materials providing specific functions such as waterproofing, thermal insulation, friction resistance, acoustic insulation, durability, decorating.

1.3 Mechanical Properties

Civil engineering materials in structures are subjected to different loads and therefore should have different mechanical properties. Some materials should have specific functions such as good waterproofing, thermal insulation, sound absorption, and adhesive properties. Due to the long-term exposure to the environment, materials also need to withstand the sunlight, rain, aging, corrosion, freezing, and other destructive effects. The mechanical properties of civil engineering materials generally include deformation, strength, elasticity, plasticity, toughness, brittleness, stiffness, ductility, hardness, viscosity, and viscoelasticity.

1.3.1 Deformation

Mechanical properties are fundamental for many construction materials, especially structural materials. Materials deform in response to loads or forces. The five basic deformations of materials include tension, compression, bending, torsion, and shear as shown in Figure 1.4.

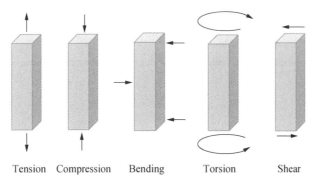

Figure 1.4 The five basic deformations of materials

1) Tension and compression

There are many types of structural members such as the steel ties of roof trusses and the suspension rods of suspension bridges. The force acting on those members is mainly tension or compression, and the main deformation of the stressed members is elongation or shortening, as shown in Figure 1.5 and Figure 1.6. Materials of these members must meet the requirements of tensile and compressive strength properties to insure sufficient reliability.

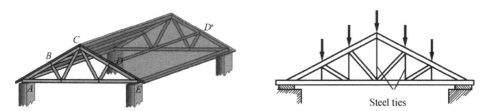

Figure 1.5 Trusses under tension

Civil Engineering Materials

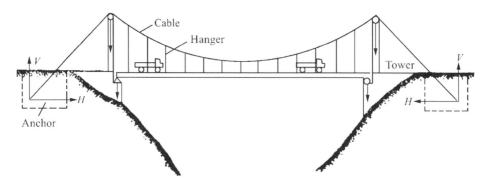

Figure 1.6 Suspension bridge elements under tension

2) Bending

Bending is the behavior of a slender structural element subjected to an external load applied perpendicularly to a longitudinal axis of the element. As shown in Figure 1.7, bending results from a couple, or a bending moment M. In pure bending, there is a neutral axis within the material where the stress and strain are zero. The beam in which the length is considerably longer than the width and the thickness is a common structural member used in buildings and bridges. Slab bending is also the key for structures, cement concrete pavements, and foundations.

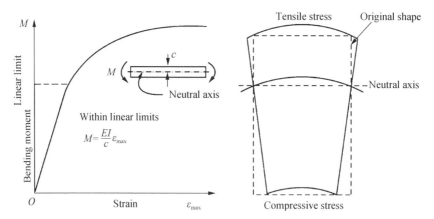

Figure 1.7 Bending of a structural element

3) Torsion

Torsion is the twisting of an object when it is loaded by a couple that produces rotation around the longitudinal axis. Torsion generates shear stress and strain. A pile or shaft can be loaded to failure more easily by torsional loads than by compressive loads. Torsional loads often control the foundation dimensions.

4) Shear

Shear deformation is a deformation of a material in which parallel internal surfaces slide past one another. Shear failure occurs when the shear force exceeds the shear capacity of different materials in the structural member.

1.3.2 Stress and Strain

Materials deform in response to loads or forces. In 1678, Robert Hooke published the first findings that documented a linear relationship between the amount of force applied to a member and its deformation. The amount of deformation is proportional to the properties of the material and its dimensions.

Stress is to divide the force by the original area. The tensile stress tends to stretch the material, as illustrated in Figure 1.8(a). It is perpendicular to the surface. It is the ratio of tensile force F_t over the original area of the cross-section A. The compressive stress can be regarded as the negative tensile stress and tends to compress the material. The shear stress tends to shear the material. It is parallel to the surface. It is the ratio of shear force F_s over the original area of the cross-section A, as illustrated in Figure 1.8(b).

Strain is to divide the deformation by the original length and therefore it is dimensionless. Tensile strain is the ratio of tensile deformation δ to the original length L_0. Lateral strain is the transverse deformation δ_L to the original width W_0, as illustrated in Figure 1.8(c). Poisson's ratio is the ratio of lateral strain to tensile strain in the direction of the stretching force. Figure 1.8(d) shows the shear strain, which is the ratio of shear deformation Δx to the original length.

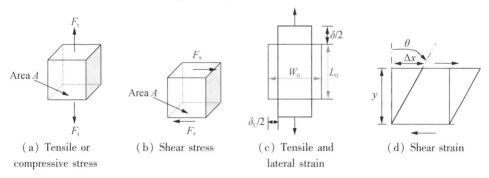

(a) Tensile or compressive stress (b) Shear stress (c) Tensile and lateral strain (d) Shear strain

Figure 1.8 Definitions of stress and strain

1.3.3 Strength

Strength is the ability of materials to resist external stresses. Figure 1.9 shows the typical stress-strain curve of a steel bar in a tensile strength test. Different types of strength can be obtained, including yield strength, ultimate tensile strength, and fracture strength. The yield strength is the maximum stress that a material can withstand without undergoing permanent or plastic deformation. The ultimate tensile strength is defined as the maximum stress that a material can withstand before failure. The fracture strength is the value of the stress at the point of fracture. As the load increases, the steel bar also experiences reversible elastic deformation and irreversible plastic deformation.

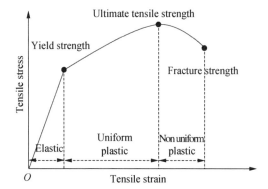

Figure 1.9 The typical stress-strain curve of a steel bar in a tensile strength test

1.3.4 Elasticity

Elasticity is the ability to resist a distorting influence or deforming force and to return to its original size and shape when that influence or force is removed. As shown in Figure 1.10, if the relationship between stress and strain is linear, it is called linear elastic

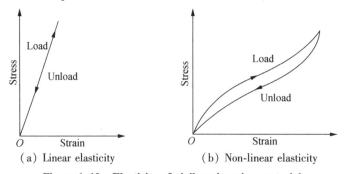

Figure 1.10 Elasticity of civil engineering materials

material. If the relationship between stress and strain is nonlinear, it is called non-linear elastic material. For example, cement concrete exhibits linear elasticity when subjected to stress for a short time within a specific range. Asphalt mixture exhibits good elastic properties at low temperatures.

1.3.5 Plasticity

Plasticity is the deformation of a material undergoing non-reversible changes of shape in response to applied forces. The deformation which does not disappear with the removal of the external force is called plastic deformation. As shown in Figure 1.11, materials show different deformation characteristics at different stages of external force action. Usually, materials show elasticity at lower stress levels. The materials exhibit plasticity or fracture characteristics after reaching the elastic limit.

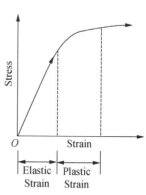

Figure 1.11 Stress-strain curves of materials in elastic and plastic stages

1.3.6 Toughness

Toughness is the ability of materials to absorb energy and deform plastically without fracturing. Toughness can be calculated as the area under the stress-strain curve, as shown in Figure 1.12(a). This value is simply called "material toughness" and it is the energy per unit volume. As shown in Figure 1.12(b), high strength does not guarantee high toughness. If the deformation before fracture is very little, the material is brittle and the toughness is low.

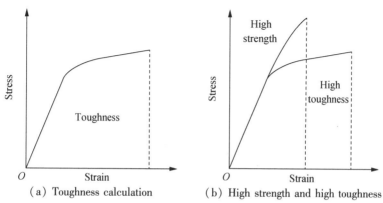

Figure 1.12 The toughness of materials

1.3.7 Brittleness

Brittleness is the phenomenon that the material breaks with little elastic deformation and without significant plastic deformation when subjected to stress. Brittleness is the opposite of toughness. As shown in Figure 1.13, the high toughness material can have large deformation before fracture, while the brittle material cannot have large deformation, and therefore its toughness is very low. Metal usually exhibits high brittleness at low temperatures.

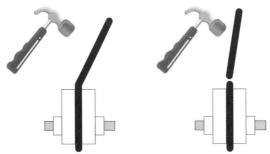

Figure 1.13 Steel bars with different toughness or brittleness

1.3.8 Stiffness

Stiffness is the resistance to elastic deformation. In the stress-strain curve, the slope or modulus can be regarded as stiffness. Stiffness reflects the relationship between the deformation of a structure and the magnitude of the force. It is used to measure the deformation of a structure in response to a force. As an example, the ratio of the tensile force to the elongation of the spring is the stiffness of the spring (Figure 1.14). Compliance or flexibility is the inverse of stiffness.
In rheology, it can be defined as the ratio of strain over stress.

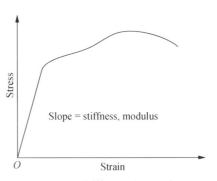

Figure 1.14 Stiffness of materials

1.3.9 Ductility

Ductility is the ability of a material to deform under tensile stress. As shown in Figure 1.15(a), the ductility of asphalt is measured by the stretched distance in centimeter at breaking. The asphalt with a very low ductility has poor cohesiveness. Figure 1.15(b) shows metals with different ductility. Metal A has low ductility and is a brittle fracture. Metal B is a ductile fracture and Metal C is a completely ductile fracture, indicating very high ductility. The most ductile metal is gold, which can be drawn into a mono-atomic wire, and then stretched about twice before it breaks.

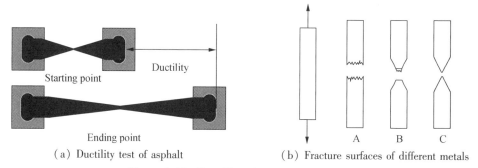

(a) Ductility test of asphalt   (b) Fracture surfaces of different metals

Figure 1.15 Ductility of asphalt and metals

1.3.10 Hardness

Hardness is the resistance to permanent shape change when a compressive force is applied. There are three types of hardness tests: scratch hardness measures how a sample resists permanent plastic deformation due to friction from a sharp object; indentation hardness measures the ability to withstand surface indentation or localized plastic deformation and the resistance of a sample to material deformation due to a constant compression load from a sharp object, as shown in Figure 1.16; rebound hardness, also known as dynamic hardness, measures the height of the "bounce" of a hammer dropped from a fixed height onto a material. Indentation hardness is usually used to distinguish between the hard and soft metals. Dynamic hardness is also used to test the relative strength of concrete.

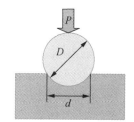

Figure 1.16 The indentation hardness test

1.3.11 Viscosity

Viscosity is the resistance to deformation at a given loading rate, indicating the internal friction of a fluid. As shown in Figure 1.17, there are two types of viscosity. The dynamic viscosity is a measurement of how difficult it is for a fluid to flow. The kinematic viscosity is the ratio of dynamic viscosity divided by the density of the liquid at the temperature of measurement. For asphalt, kinematic viscosity is usually measured with a capillary viscometer and the dynamic viscosity can also be calculated.

Figure 1.17　The viscosity of asphalt

1.3.12　Viscoelasticity

Viscoelasticity means materials exhibit both viscous and elastic characteristics when undergoing deformation. It is a very important property of asphalt and asphalt concrete. The strain is an immediate response to stress for elastic and elastoplastic materials, but viscoelastic materials have a delayed response to the load. Figure 1.18 shows a sinusoidal axial load applied on a viscoelastic material such as asphalt concrete, versus time. Also, it shows the resulting deformation versus time, where the deformation lags the load, i.e. the maximum deformation of the sample occurs after the maximum load is applied. The amount of time delay of the deformation depends on the material characteristics and the temperature.

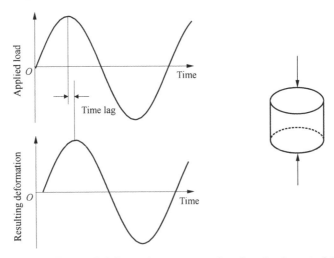

Figure 1.18　Load-deformation response of a viscoelastic material

1.4 Non-mechanical Properties

Non-mechanical properties refer to other characteristics of materials other than load response, which influence the selection, use, and performance of materials. Several types of properties that are of interest to civil engineers include density, thermal properties, surface characteristics, water-related properties, durability, and workability.

1.4.1 Density

There are different definitions of density depending on the volume or weight considered due to the permeable voids and non-permeable voids in the material. As shown in Figure 1.19, the bulk volume of the material, denoted as V_0, includes three parts: the volume of the solid, denoted as V; the volume of non-permeable voids, denoted as V_K; the volume of permeable voids, denoted as V_B. Air voids, denoted as V_P, include both non-permeable and permeable voids. The apparent volume, denoted as V_a, includes the volume of the solid V and the non-permeable voids V_K.

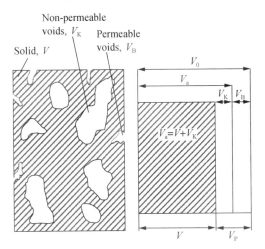

Figure 1.19 Definitions of different volumes

To measure the solid density, we can grind the material into powder to eliminate the non-permeable voids. For example, the maximum theoretical density of an asphalt mixture is measured with the "rice samples". The solid density of the material is the ratio of the weight over the volume of the solid.

Civil Engineering Materials

$$\rho = \frac{m}{V} \tag{1.1}$$

where ρ = solid density (g/cm^3);
  m = weight of the solid (g);
  V = volume of the solid (cm^3).

Bulk density is the ratio of the weight of solid over the bulk volume of the material.

$$\rho_0 = \frac{m}{V_0} \tag{1.2}$$

where ρ_0 = bulk density (g/cm^3);
  V_0 = bulk volume of the material (cm^3).

Apparent density is the ratio of the weight of solid over the apparent volume of the material.

$$\rho_a = \frac{m}{V + V_K} \tag{1.3}$$

where ρ_a = apparent density (g/cm^3);
  V_K = volume of non-permeable voids (cm^3).

Porosity is the ratio of the volume of voids over the bulk volume. It also equals 1 minus the ratio of bulk density over solid density.

$$P_0 = \frac{V_0 - V}{V_0} = 1 - \frac{\rho_0}{\rho} \tag{1.4}$$

where P_0 = porosity $(\%)$.

For material particles packed in the container, in addition to the solid, non-permeable, and permeable voids, there are voids between particles, V_J, as shown in Figure 1.20. Unit weight is the ratio of the weight of solid over the volume of the container. It is needed for the proportioning of mixes or to estimate the weight of materials during purchasement and transportation.

$$\rho_0' = \frac{m}{V_0'} \tag{1.5}$$

where ρ_0' = unit weight (g/cm^3);
  V_0' = volume of the container (cm^3).

Void fraction is the ratio of the voids between particles, which is the volume of the voids between particles over the volume of the container. It equals one minus the ratio of unit weight to bulk density.

$$P_0' = \frac{V_0' - V_0}{V_0'} = 1 - \frac{\rho_0'}{\rho_0} \tag{1.6}$$

where P'_0 = void fraction (%).

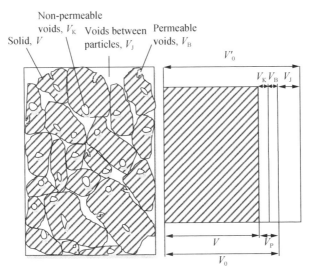

Figure 1.20 Definitions of different volumes in the container

1.4.2 Thermal Properties

Thermal properties are not only critical to determine the temperature-induced stress and strain, but also are important for some functions such as thermal insulation. The thermal properties of civil engineering materials generally include thermal conductivity, heat capacity, specific heat capacity, thermal expansion, etc.

1) Thermal conductivity

The ability of a material to conduct heat is called thermal conductivity. Thermal conductivity is numerically equal to the amount of heat that passes through a unit area (1 m²) per unit time (1 s) of a material with a thickness of 1 m when the temperature difference between its relative surfaces is 1 K. It can be expressed in Equation (1.7).

$$\lambda = \frac{Q\delta}{At(T_2 - T_1)} \tag{1.7}$$

where λ = thermal conductivity (W/(m·K));
  Q = heat transferred (J);
  A = heat transfer area (m²);
  t = heat transfer time (s);
  δ = thickness of the material (m);

$T_2 - T_1$ = temperature difference between the two sides of the material (K).

The thermal conductivity of various civil engineering materials varies greatly, roughly between 0.035 and 3.5 W/(m · K). The thermal conductivity is closely related to the pore structure of the material. Since the thermal conductivity of confined air is very small, around 0.023 W/(m · K), the material with high porosity usually has low thermal conductivity. The thermal conductivity of the material will be greatly increased when it is exposed to moisture or freezing. This is because the thermal conductivity of water (0.58 W/(m · K)) and ice (2.20 W/(m · K)) is much higher than that of air. Therefore, insulation materials should always be dry to facilitate their thermal performance.

2) Heat capacity

The property of a material to absorb heat when it is heated and to release heat when it is cooled is called heat capacity. The magnitude of heat capacity is expressed in terms of specific heat capacity, also called heat capacity coefficient or specific heat. Specific heat capacity is the amount of heat absorbed when the temperature of a material rises by 1 K or the amount of heat given off when the temperature decreases by 1 K. The amount of heat absorbed or given off by a material can be calculated by Equations (1.8) and (1.9).

$$Q = cm(T_2 - T_1) \qquad (1.8)$$

$$c = \frac{Q}{m(T_2 - T_1)} \qquad (1.9)$$

where Q = heat absorbed or given off by the material (J);

c = specific heat capacity of the material (J/(g · K));

m = weight of materials (g);

$T_2 - T_1$ = temperature difference before and after the material is heated or cooled (K).

Specific heat is a physical quantity that reflects the magnitude of a material's ability to absorb or release heat. Different materials have different specific heat. The same material can have different specific heat at different states of matter. For example, the specific heat of water is 4.186 J/(g · K), while the specific heat of ice is 2.093 J/(g · K).

The specific heat of the material influences the internal temperature stability of the structure. Materials with high specific heat can moderate the temperature fluctuations in the room when the heat flow changes or when the heating equipment is not uniformly heated. The specific heat of commonly used civil engineering materials is shown in Table 1.1.

1 Fundamentals of Civil Engineering Materials

Table 1.1 Thermal property index of several typical materials

Items	Steel	Concrete	Pinewood	Brick	Granite	Airtight air	Water
Specific heat capacity (J/(g · K))	0.48	0.84	2.72	0.88	0.92	1.00	4.18
Thermal conductivity (W/(m · K))	58	1.51	0.12	1.31	3.49	0.023	0.58

3) Thermal insulation

Thermal insulation is the reduction of heat transfer between objects in thermal contact or the range of radiative influence. For civil engineering structures, it includes the reduction of the loss of heat and the entry of heat from outside. The smaller the thermal conductivity of a material, the greater the thermal resistance. Buildings with good thermal insulation are energy efficient to keep warm in the winter or cool in the summer.

4) Thermal expansion

The coefficient of thermal expansion is the amount of expansion per unit length due to one unit of temperature increase. The liner and volumetric coefficient of thermal expansion can be calculated by Equations (1.10) and (1.11).

$$\alpha_L = \frac{\delta_L}{\delta_T \times L} \tag{1.10}$$

$$\alpha_V = \frac{\delta_V}{\delta_T \times V} \tag{1.11}$$

where α_L = linear coefficient of thermal expansion (1/°C);

α_V = volumetric coefficient of thermal expansion (1/°C), for isotropic materials, $\alpha_V = 3\alpha_L$;

δ_T = change of temperature (°C);

δ_L = length change of the specimen (cm);

L = original length of the specimen (cm);

δ_V = volume change of the specimen (cm^3);

V = original volume of the specimen (cm^3).

The coefficient of thermal expansion is very important in the design of structures. Generally, structures are composed of many materials that are bound together. The material with less expansion will restrict the straining of other materials. This constraining effect will cause stress that may lead to fracture. Stress can also be developed as a result of a thermal gradient in the structure. When the structure is restrained, stress develops in the material. Concrete pavements that are restrained from movement, may crack in the winter due to a drop in temperature and may "blow up" in the summer due to an increase in temperature. Joints are therefore used in buildings, bridges, concrete pavements, and various structures to accommodate this thermal effect.

1.4.3 Surface Properties

The surface properties of civil engineering materials mainly include corrosion and degradation, abrasion and wear resistance, friction, and the surface texture.

1) Corrosion and degradation

Nearly all materials deteriorate over their service lives. Corrosion or degradation involves the deterioration of the material when exposed to an environment resulting in the loss or performance reduction of the material. Metal corrosion involves oxidation-reduction reactions in which the metal is lost by dissolution at the anode. Asphalt ages due to oxidation, ultraviolet radiation, etc. The protection of materials from environmental corrosion is an important design concern. Coating or improving the surface impermeability is an effective way to improve the corrosion resistance of materials.

2) Abrasion and wear resistance

Abrasion and wear resistance is to prevent material loss or deterioration from surface abrasion or wearing. For example, pavement surface must have sufficient resistance to the wearing and polishing of vehicle tires to provide adequate skid resistance for vehicle braking and turning. The specifications, AASHTO T 96-02 or ASTM CBI/CBIM-20, provide requirements on the abrasion and wear resistance of aggregates used in asphalt mixtures and cement concrete.

3) Friction

Friction is the resistance to motion of one object moving relative to another. It is an important property for the pavement surface because pavement friction gives drivers the ability to control the vehicles in a safe manner, in both longitudinal and lateral directions. The higher friction on the pavement surface, the more control the driver has over the vehicle. It is a key for highway geometric design as well as the material design for the pavement surface course. Pavement friction can be measured by the pendulum tester, the locked-wheel skid tester, or the surface texture.

4) Surface texture

Different surface textures are required for different materials. A smooth texture of aggregate particles can improve the workability of fresh Portland cement concrete. In contrast, a rough texture of aggregate particles is needed in asphalt concrete mixtures to provide a strong skeleton to resist permanent deformation under the wheel loads. In addition, for pavement surfaces, a rough surface texture can provide adequate friction resistance.

1.4.4 Water-related Properties

Water is a critical factor for many structures including hydraulic structures, bridges, dams, pavements, and foundations. Water is an important ingredient in all hydraulic cementing materials while can cause durability problems if not properly treated. Water-related properties generally include affinity with water, permeability, water absorption, moisture susceptibility, and freeze-thaw resistance.

1) Affinity with water

Materials can be classified into hydrophobic and hydrophilic, as shown in Figure 1.21. Hydrophobic material tends to repel or fail to mix with water. Its contact angle θ is no less than 90°. Hydrophilic material tends to mix with or be wetted by water. Its contact angle θ is no more than 90°. Most civil engineering materials, such as stone, aggregate, brick, concrete, wood, etc., are hydrophilic materials. The surface of these materials can be wetted by water, and water can be absorbed inside the capillary tube of the material through capillary action. Asphalt and polymers are hydrophobic materials whose surfaces are water-repellent. These materials prevent water from penetrating the capillaries, thus reducing the water absorption of the material.

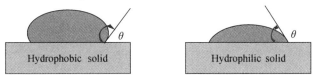

Figure 1.21 Different affinity with water

2) Permeability

Permeability is the capacity of a porous material for transmitting a fluid. Permeability coefficient measures the volume of a fluid that flows in unit time through a unit volume of the material. Figure 1.22 shows an apparatus for testing the permeability coefficient for porous concrete or the asphalt mixture.

$$K = \frac{Wd}{Ath} \quad (1.12)$$

where K = permeability coefficient (cm/s);
W = volume of water passing through the specimen (cm³);
t = time to pass through the specimen (s);
A = cross-sectional area of the specimen (cm²);
h = constant head causing flow (cm);
d = thickness of the specimen (cm).

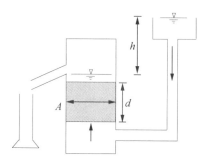

Figure 1.22 A typical apparatus for testing permeability coefficient

3) Water absorption

The ability of a material to absorb water when immersed in water is called water absorption. The water absorption of a material is not only related to the hydrophilicity or hydrophobicity of the material, but also the size and the characteristics of the permeable pores. Usually, the greater the porosity of the material, the higher the absorption.

Figure 1.23 shows the voids and moisture condition of aggregates. The bone-dry condition means no water. The air-dry condition means a little water inside the permeable void. The saturated surface-dry condition means the permeable void is filled with water. The moist condition means extra water films on the surface. Absorption by weight is the ratio of the weight of water over the dry weight of the material. Absorption by volume is the ratio of the volume of water in the materials over the dry volume.

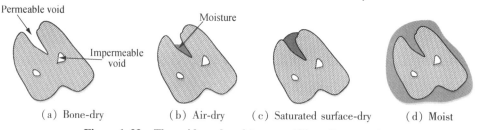

Figure 1.23 The voids and moisture condition of aggregates

Water absorption can be measured by the ratio of the weight or volume of absorbed water to the dry weight or volume of the material. The weight absorption of some lightweight materials, such as the lightweight aggregates, can be higher than 100%. In this case, volumetric water absorption is often used. Usually, the presence of water in material harms its properties, including increasing the volume, apparent density, and thermal conductivity, reducing the strength, and causing freeze-thaw risks.

4) Moisture susceptibility

Moisture susceptibility is usually calculated as the ratio of performance such as strength after moisture conditioning over the performance before conditioning. Moisture susceptibility is a complex phenomenon dependent upon many factors, including absorption, porosity, strength, etc. For asphalt mixtures, the moisture susceptibility is influenced by the moisture content in the asphalt mixture and the adhesion of the asphalt binder to the aggregate surface.

5) Freeze-thaw resistance

Freeze-thaw resistance is the ratio of performance, such as strength after freeze-thaw cycles over the performance before freeze-thaw cycles. Freeze-thaw resistance is required for both cement concrete and asphalt mixtures. During freeze-thaw conditioning, the water is continuously absorbed into the pores and micro-cracks of the concrete or mixtures. When water freezes and expands at low temperatures, it produces tremendous pressure on the pores of the concrete or mixtures. The accumulative effect of freeze-thaw cycles can eventually cause cracking, spalling, and scaling of the materials.

1.4.5 Durability

Civil engineering materials in general, such as stone, masonry, cement concrete, asphalt concrete, etc., are all constantly subject to the physical and chemical actions of the environment they are exposed to. Freeze-thaw and abrasion are mainly physical processes while others mainly include chemical reactions. When materials are in the water or a moist environment, they are subject to the risk of the chemical or physical attack of the water. Asphalt and polymers are aging when exposed to sunlight, air, and moisture. Durability is the resistance of materials to environmental corrosion including physical processes and chemical reactions. The concrete durability includes freeze-thaw resistance, abrasion resistance, corrosion, alkali reaction, carbonation, etc. The durability of asphalt mainly refers to aging which includes complicated physical and chemical changes. The durability of steel is mainly rusting. To improve the durability, corresponding measures can be taken according to the environment and material characteristics. The destructive effects of the surrounding media can be reduced by coating, insulation and excluding aggressive substances, etc. The resistance of the material itself can also be improved by increasing the impermeability, eliminating potentially risky compositions.

1.4.6 Workability

Even if a material is well suited to a specific application, production and construction considerations may not permit the use of the material. Production considerations include the availability of the material and the ability to fabricate the material into the desired shapes. Construction considerations address all the factors that relate to the ability to build, cast, or pave the structure on site. For example, cement concrete requires good flowability, cohesiveness, and water retention. Usually, water-reducing admixtures are added to improve the workability of concrete or to produce high-strength or quick-set concrete. Asphalt mixtures also need to be heated to a specific range of temperature for mixing and paving. For asphalt pavement pothole patching, rapid-curing cutback asphalt mix is usually adopted.

1.5 Variability

Variability is an important indicator of the quality of civil engineering materials. Less variability usually indicates better quality. For example, the quality and quantity of water, cement, aggregates, and admixtures all influence the performance of cement concrete. Usually, a specific number of samples are required to be tested to obtain the results. The test results still include variability which is influenced by three sources: the inherent variability of the material, the variance caused by the sampling, and the variance related to the test methods. Therefore, the design of experiment (DOE) and statistical significance tests are usually conducted to investigate the effects of material types and proportions.

The definitions of precision and accuracy are fundamental to understanding the variability. As shown in Figure 1.24, precision is the closeness of the measurements to each other while accuracy is the closeness of the measurements to a specific value. The deviation of the measurements to the true value is the bias or errors. The errors can be classified as systematic errors and human errors. We want to eliminate systematic errors and reduce human errors. Figure 1.24(a) is precise but not accurate. Although the measurements are close to each other, meaning high precision, they are far away from the true value. Figure 1.24(b) is accurate because the average of the measurements is close to the true value, but not precise because of the high variability. Figure 1.24(c) is both accurate and precise.

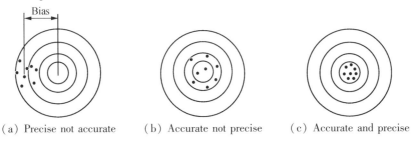

(a) Precise not accurate   (b) Accurate not precise   (c) Accurate and precise

Figure 1.24 Precision and accuracy

Questions

1. Define the following material deformations.
(1) Tension   (2) Compression   (3) Bending
(4) Torsion   (5) Shear

2. Define the stress and strain due to tension, compression, and shear.

3. Briefly discuss the yield strength, ultimate tensile strength, and fracture strength of a steel base in the tensile strength test.

4. What is the difference between toughness, brittleness, stiffness, and ductility?

5. How to measure the hardness?

6. Explain the elasticity, plasticity, viscosity, and viscoelasticity of materials.

7. What are the common parameters used to evaluate the thermal properties of materials?

8. Discuss the differences between density, apparent density, and bulk density.

9. Discuss why the water affinity of materials needs to be considered in civil engineering.

10. Briefly discuss the variability of construction materials. Define the terms of accuracy and precision when tests on materials are performed.

2 Aggregates

Aggregates are intensively used in civil engineering and transport infrastructures. In specification ASTM C125-21, aggregates are defined as granular materials such as sand, gravel, and crushed stones or slag, used with a cementing medium to form hydraulic cement concrete or mortar. In specification JTG E42—2005, aggregates are defined as granular materials that are used in mixture as skeleton structure and filling, including gravel, crushed stones, crushed sand, stone chips, sand, etc.

Aggregates can be used either as underlying materials for foundations and pavements or as ingredients in Portland cement and asphalt concrete. As underlying materials, aggregates can provide a stable base or a drainage layer. As ingredients, aggregates can provide strong inter-particle friction, which is the key to the strength and stability of concrete. Aggregates account for 75 %– 85 % by weight in Portland cement concrete, and 92 %– 96 % by weight in the asphalt mixtures. This chapter introduces the classification, production, physical and mechanical properties of coarse and fine aggregates, and gradation design.

2.1 Production of Aggregates

2.1.1 Aggregate Sources

According to the sources, aggregates can be classified as natural aggregates, manufactured aggregates, and recycled aggregates. Natural aggregates include gravel which is obtained from gravel pits and river deposits, and crushed stones which are obtained by processing rocks from quarries. It is noted that gravel also needs to be crushed to obtain the required size and shape. The manufactured aggregates include the by-product aggregates which are mostly the slag waste from iron and steel production and the expanded shale and clay which are produced by the drying and sintering of clay or shale raw material, forming a sintered porous structure. Recycled aggregates are obtained from the demolition of buildings, bridges, and other structures. Generally, recycled concrete aggregates contain not only the original aggregates, but also the cement paste or asphalt mortar, which may reduce the specific gravity and increases the porosity.

According to the density, aggregates can be classified into heavyweight, normalweight, and lightweight aggregates. The traditional concrete using natural aggregates is normalweight concrete. Slag aggregates are heavyweight aggregates and are usually used in the heavyweight concrete for radiation shields. Expanded shale, clay, or slate materials that have been calcined in a rotary kiln to develop porous structures are lightweight aggregates and can be used in lightweight concrete for heat or sound insulation.

2.1.2 Types of Rocks

Rock has been an important civil engineering material since the beginning of civilization. Compared to other abundant natural materials such as timber, rock is much harder and more durable whereas very heavy. Rocks constitute the earth's crust and are mainly crystalline solids with definite physical and chemical properties which also determine the properties of aggregates. Rocks can be classified into three types based on how they are formed, i. e. igneous, sedimentary, and metamorphic rocks.

1) Igneous rocks

Igneous rocks are hardened or crystallized volcanic molten materials. There are two types of igneous rocks: the extrusive or volcanic rocks formed by the rapid cooling of the magma at the earth's surface and the intrusive rocks formed by the slow cooling of the magma at depths within the earth's crust. The most frequently seen igneous rocks include granite, basalt, diabase, pozzolan, and pumice. Granite, basalt, and diabase have higher density and strength while pozzolan and pumice have much lower density and strength.

(1) Granite

The composition and texture of rocks can be characterized by petrographic analysis. Granite is an intrusive igneous rock showing a variety of tonalities (white, grey, blue, pink, or red) and a holocrystalline granular texture ranging from less than 0.1 mm to 10 cm. Granite is mainly composed of feldspar and quartz. It usually contains 20% to 60% quartz by volume and is hard and usually difficult to be mechanically processed. Based on the chemical composition, granite is typically described as the ultrabasic rock with silica oxides (SiO_2) content higher than 40% by weight. The common types of granite contain more than 60% silica oxides. Other compounds of granite include alumina (Al_2O_3), iron oxide Ⅲ (Fe_2O_3), magnesium oxide (MgO), sodium oxide (Na_2O), and potassium oxide (K_2O). The apparent density of granite is between 2.65 and 2.75 g/cm^3 and the compressive strength usually lies above 200 MPa.

(2) Basalt and Diabase

Basalt is mostly formed as an extrusive rock with dark color and pyroxene and olivine crystals. Basalt usually includes 45% to 55% silica oxides and less than 10% feldspar. Its apparent density is around 3 g/cm^3 and the compressive strength is between 100 and 500 MPa. Diabase is similar to basalt and normally has finer crystals. Its apparent density is between 2.7 and 2.9 g/cm^3 and the compressive strength is between 160 and 180 MPa. It has good resistance to acid.

(3) Pozzolan and Pumice

Pozzolan is a powdery volcanic rock with a grain size of less than 5 mm. It has a pozzolanic activity that can react with calcium oxide and therefore can be used as raw materials for the supplementary materials of Portland cement. Pumice is a volcanic rock with a grain size of larger than 5 mm and a porous structure. Its apparent density is between 0.3 and 0.6 g/cm^3 which is less than the density of water.

2) Sedimentary rocks

Sedimentary rocks are deposits of disintegrated existing rocks, soils, or other inorganic materials. Natural cementing binds the particles together. Sedimentary rocks can be classified by the predominant mineral, including calcareous (limestone, chalk, etc.), siliceous (chert, sandstone, etc.), and argillaceous (shale, etc.).

(1) Limestone

Limestone is the most widely used sedimentary rock. Its major components are calcite and aragonite, which are different crystal forms of calcium carbonate. Limestone can be classified according to the mode of formation, and show many colors including white, grey, buff, and blue. Its apparent density is 2.6 to 2.8 g/cm^3. The compressive strength is 80 to 160 MPa. Its water absorption is between 2% and 10%.

(2) Sandstone

Sandstone is composed of sand-sized mineral particles or rock fragments. Depending on the original sand deposit, the sandstone has a texture ranging from 0.06 to

2 mm. Sandstone can have different colors including white, buff, grey, brown, and shades of red depending on the natural cement. Its apparent density is around 2.65 g/cm³, the compressive strength ranges from 5 to 200 MPa and its water absorption is between 0.2% and 7%.

3) Metamorphic rocks

When igneous or sedimentary rocks are drawn back to the earth's crust and exposed to heat and pressure, then the grain structure is reformed and the metamorphic rocks are formed. Metamorphic rocks generally have a crystalline structure, with grain sizes ranging from fine to coarse. Metamorphic rocks frequently have anisotropic texture and typical examples are gneiss, and marble. Marble is a metamorphic stone with higher hardness, formed at a variable depth from the earth's surface. Marble forms from complete recrystallization of other rocks. It is mainly used for building decoration. The compressive strength is between 50 and 140 MPa.

2.1.3 Physical Properties of Rocks

Natural aggregates are produced from the breakdown of large rocks. The properties of aggregates are mostly determined by the rocks and the processing method. The technical properties of rocks include physical, mechanical, and chemical properties.

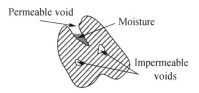

Figure 2.1 Two types of voids in a rock particle

Physical properties of rocks generally include density, absorption, and permeability. They are directly related to the physical properties of aggregates and are often used as performance indices. To calculate the density of rocks, we first need to understand the two types of voids in a rock particle as shown in Figure 2.1. The impermeable voids are the pores inside the rock particle, whereas the permeable voids or open pores are connected to the surface and can absorb moisture or be filled with water.

(1) Solid Density: The solid density in rocks is obtained by measuring the crushed rock particles which do not include the permeable and impermeable air voids.

(2) Bulk Density: The bulk density is the ratio of the weight of rock over the bulk volume which includes the permeable voids.

(3) Apparent Density: The apparent density is the ratio of the weight of rock over the apparent volume which excludes the permeable voids.

(4) Moisture Content: Moisture content means the water content in the air dry state. It is the ratio of the weight of water to the dry weight of the rock particles.

(5) Absorption at Saturation: The absorption at saturation means the maximum water content when all the permeable voids of rock particles are filled with water, which is also called the saturated surface dry (SSD) status.

1) Water damage resistance

The water damage resistance is evaluated by the softening coefficient. A higher softening coefficient means higher resistance to water damage. The water damage resistance of rocks can be classified into four levels based on the softening coefficient.

- High: >0.9
- Medium: 0.75 – 0.9
- Low: 0.6 – 0.75
- Unstable: <0.6

2) Freeze-thaw resistance

Freeze-thaw resistance is evaluated by the reduction in strength or weight loss after specific freeze-thaw cycles. As shown in Figure 2.2, when the absorbed water in the permeable voids freezes, its volume expands and causes cracks or forces cracks to widen. During thawing, more water may be absorbed into the cracks and the fracture action continues. A quality rock should have a strength reduction less than 25 % and a weight loss less than 5 %, and have no penetrating cracks.

Figure 2.2 Cracks caused by freezing

3) Soundness

The soundness is tested by submerging the rock sample in a saturated solution of sodium sulfate and letting salt crystals form in permeable pores, thus simulating the formation of ice crystal. The damage caused by the expanded crystals is observed to evaluate the soundness.

2.1.4 Mechanical Properties of Rocks

Mechanical properties of rocks include compressive strength, impact toughness, and abrasion resistance and are used to evaluate the ability to withstand loads, wearing, or deformation.

(1) Compressive Strength: The compressive strength of rocks is measured by the

unconfined compressive test of the core of rock samples. It measures the ability of rock samples to withstand compressive loads and resist deformation.

(2) Impact Toughness: Impact toughness measures the rock fracture toughness under impacts. It measures the ability of rock samples to absorb energy and resist fracturing when force is applied.

(3) Abrasion Resistance: Abrasion resistance is the behavior of rock samples under the shear stress and gouging action of abrasion. It indicates the wearing quality when the rock is exposed to traffic loading.

2.1.5 Chemical Properties of Rocks

The chemical properties of rocks related to aggregates mainly include the alkali reaction and affinity with water or asphalt. The result of the alkali reaction test indicates whether the rock is inert or active, which is critical for the alkali reaction in cement concrete. The affinity with water or asphalt indicates the bonding between aggregates and asphalt. Usually, limestone has a better affinity with asphalt than granite.

2.1.6 Crusher

For aggregates using natural resources, rocks are usually obtained from quarries and crushed to obtain the needed size, shape, and texture. There are generally three types of crushers: the jaw crusher, the cone crusher and the impact crusher. As shown in Figure 2.3, the jaw or cone crusher uses a compressive force to crush aggregates, while the impact crusher uses an impact force to crush aggregates. In the impact crusher, aggregates are firstly crushed by the rotating bars and then thrown against the breaker plates for secondary crushing. The jaw or cone crusher tends to produce more aggregates with sharp angles due to the shear fracture of aggregates under the compressive or squeezing force.

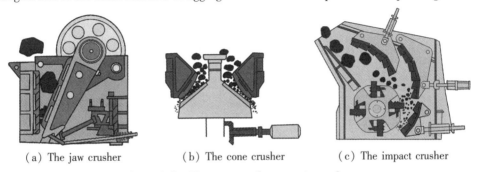

(a) The jaw crusher   (b) The cone crusher   (c) The impact crusher

Figure 2.3 Three types of aggregate crushers

The impact crusher tends to produce more round shape aggregates. Usually, the jaw or cone crusher is used to crush larger rocks and the impact crusher is then used to crush aggregates into smaller sizes.

2.2 Physical Properties of Coarse Aggregates

The properties of aggregates are defined by the characteristics of both the individual particles and the combined materials. In specification JTG E42—2005, the pavement aggregate tests include 26 tests for coarse aggregates, 19 tests for fine aggregates, and 5 tests for mineral fillers, covering the physical, mechanical and chemical characterizations of aggregates. Table 2.1 summarize the quality requirements for coarse aggregates used in asphalt mixtures and corresponding test methods from JTG E42—2005, mainly including the physical and mechanical properties.

Table 2.1 Requirements for coarse aggregates in asphalt mixtures

Test items	Expressway and Class I road		Other roads	Test methods
	Surface	Other layers		
Max. crushing value (%)	26	28	⩽30	T 0316
Max. LA abrasion value (%)	28	30	35	T 0317
Min. apparent specific gravity	2.60	2.50	2.45	T 0304
Max. absorption (%)	2.0	3.0	3.0	T 0304
Max. soundness (%)	12	12	—	T 0314
Max. flakiness content (%)	15	18	20	T 0312
Max. percentage of > 9.5 mm (%)	12	15	—	
Max. percentage of <9.5 mm (%)	18	20	—	
Max. percentage of <0.075 mm (%)	1	1	1	T 0310
Max. content of soft particles (%)	3	5	5	T 0320

2.2.1 Sampling

The sample of aggregates being tested must represent the whole population of aggregates. For coarse aggregates, the sample size is determined by the nominal maximum size of aggregates. Larger-sized aggregates require larger samples to minimize errors. Usually, field samples are reduced using sample splitters or by quartering to obtain the samples needed for testing. The specification JTG E42—2005 T0301 recommends a quartering method to obtain coarse aggregate samples, which includes flattening and quartering the mixed field samples, and then retaining opposite quarters as the sample as shown in Figure 2.4.

Figure 2.4 Quartering and taking opposite quarters

2.2.2 Sieve Analysis and Gradation

Sieve analysis is to evaluate the size of aggregates by passing the aggregates through a series of sieves of progressively smaller mesh size (Figure 2.5). As defined in specification JTG E42—2005 T0302 and T0327, the common sieve sizes range from 0.075 to 75 mm, including 75, 63, 53, 37.5, 31.5, 26.5, 19, 16, 13.2, 9.5, 4.75, 2.36, 1.18, 0.6, 0.3, 0.15, and 0.075 mm. Gradation is the particle size distribution of aggregates and is one of the most important characteristics for aggregates, especially in pavement asphalt mixtures. The gradation results are usually described by the cumulative percentage of aggregates that either pass through or are retained by a specific sieve size.

Figure 2.5 Sieve analysis

As shown in Table 2.2, from sieve analysis, we can obtain the weight retained on each sieve and the pan, the total weight and the weight lost during sieve analysis, which can be calculated as below.

$$m_s = m_t - \sum m_i \tag{2.1}$$

where m_s = weight lost during sieve analysis (g);
m_t = total weight (g);
m_i = weight retained on the ith sieve and the bottom (g).

Civil Engineering Materials

Table 2.2　Results of sieve analysis

Sieve (mm)	Retained (g)	Retained (%)	Cumulative retained (%)	Passing (%)
$D_0 = D_{max}$	$m_0 = 0$	$a_0 = 0$	$A_0 = 0$	$P_0 = 100$
D_1	m_1	$a_1 = m_1 / \sum m_i \times 100$	$A_1 = a_1$	$P_1 = 100 - A_1$
D_2	m_2	$a_2 = m_2 / \sum m_i \times 100$	$A_2 = a_1 + a_2$	$P_2 = 100 - A_2$
D_3	m_3	$a_3 = m_3 / \sum m_i \times 100$	$A_3 = a_1 + a_2 + a_3$	$P_3 = 100 - A_3$
...	...	...	...	...
D_i	m_i	$a_i = m_i / \sum m_i \times 100$	$A_i = a_1 + a_2 + a_3 + \cdots + a_i$	$P_i = 100 - A_i$
...	...	...	...	...
D_n	m_n	$a_n = m_n / \sum m_i \times 100$	$A_n = a_1 + a_2 + a_3 + \cdots + a_n$	$P_n = 100 - A_n$

Based on the weight retained on each sieve, we can calculate the percentage retained which is the weight retained on each sieve over $\sum m_i$, the cumulative percentage retained which is the sum of the percentage retained above this sieve, and the passing percentage as below.

$$a_i = \frac{m_i}{\sum m_i} \times 100 \quad (2.2)$$

$$A_i = a_1 + a_2 + \cdots + a_i \quad (2.3)$$

$$P_i = 100 - A_i \quad (2.4)$$

where a_i = percentage retained (%);
　　　A_i = cumulative percentage retained (%);
　　　P_i = passing percentage (%).

For the example shown in Table 2.3, the weight retained on the 4.75 mm sieve is 1000 g and the total weight of the aggregate is 2000 g. The retained percentage on this sieve is 1000/2000 = 50%. The cumulative retained percentage can be calculated by adding the retained percentage of all the sieves above the 4.75 mm sieve, which equals 60%, indicating that 60% of the sample are larger than 4.75 mm. The passing percentage is one minus the cumulative retained percentage, which equals 40%, indicating 40% of the sample passes the 4.75 mm sieve.

Table 2.3 The example of a sieve analysis

Sieve (mm)	Retained (g)	Retained (%)	Cumulative retained (%)	Passing (%)
13.2	0	0	0	100
9.5	200	10	10	90
4.75	1 000	50	60	40
2.36	600	30	90	10
1.18	0	0	90	10
0.6	200	10	100	0
0.3	0	0	100	0
Bottom	0	0	100	0

2.2.3 Aggregate Size

Aggregates can be classified by size as coarse aggregates, fine aggregates, and fillers (Figure 2.6). As shown in Table 2.4, in the ASTM specification, coarse aggregates are larger than 4.75 mm. Fine aggregates are between 4.75 mm and 0.075 mm. Fillers are smaller than 0.075 mm. In the Chinese specification JTG E42—2005, for asphalt mixtures, the threshold for classifying coarse and fine aggregates is 2.36 mm while for ce-

Figure 2.6 Aggregates of different sizes

ment concrete, the threshold for classifying coarse and fine aggregates is 4.75 mm. Generally, large aggregates have a smaller specific surface area and require less binder, and are economical when using in mixes. However, large aggregate mixes tend to have poor workability and cause more voids.

Table 2.4 Aggregate size

Types	JTG E42—2005		ASTM
	Asphalt mixtures	Cement concrete	
Coarse	>2.36 mm	>4.75 mm	>4.75 mm
Fine	0.075 – 2.36 mm	≤4.75 mm	0.075 – 4.75 mm
Filler	<0.075 mm	—	<0.075 mm

There are two maximum sizes based on the results of the sieve analysis. The maximum aggregate size is the smallest sieve through which 100% of the aggregate sample particles pass. The maximum aggregate size cannot represent the aggregate size when the amount of large particles is very limited. Therefore, the nominal maximum aggregate size is defined, which is the largest sieve that retains some of the aggregate particles but generally not more than 10% by mass.

Table 2.5 shows the passing percentage of three groups of aggregates at different sieve sizes. For group 1, the passing percentages at 25 mm and 19 mm are 100% and 93%, respectively. According to the definition, its maximum size is 25 mm and its nominal maximum size is 19 mm. For group 2, it is easy to determine that its maximum size is 19 mm and its nominal maximum size is 16 mm. For group 3, its maximum size is 19 mm. But its passing percentage at 16 mm is 80%, indicating the retained percentage on 16 mm sieve is 20%, higher than 10%. Therefore its nominal maximum size should still be 19 mm.

Table 2.5 The passing percentage of three groups of aggregates

Samples		Group 1	Group 2	Group 3
Sieve size	25 mm	100%	100%	100%
	19 mm	93%	100%	100%
	16 mm	80%	93%	80%
Maximum size		25 mm	19 mm	19 mm
Nominal maximum size		19 mm	16 mm	19 mm

2.2.4 Moisture Condition

Aggregates can absorb water and asphalt in surface or permeable voids. The amount of water absorbed by the aggregate is important in the design of Portland cement concrete. Although the absorbed water usually cannot react with the cement or improve the workability of the fresh concrete, the aggregate absorption must be considered to determine the amount of water mixed into the concrete. For asphalt concrete, absorbed asphalt can help improve the bonding between the asphalt and the aggregate, whereas too much absorption requires a great amount of asphalt, increasing the cost of the mix. Therefore, low-absorption aggregates are preferred for asphalt concrete.

Figure 2.7 shows the four moisture conditions for an aggregate particle. In a bone-dry condition, the aggregate particle contains no moisture, which usually requires drying the aggregate in an oven. In an air-dry condition, the aggregate particle contains some moisture but is not saturated. In a saturated surface-dry (SSD) condition, the aggregate's permeable voids are filled with water but the main surface area of the aggregate particle is dry. In a moist condition, the aggregate particle has water on the surface area in excess of the SSD condition. The fifth absorption condition only occurs in the asphalt mixture, in which only a portion of the water-permeable voids of the aggregate particle is filled with asphalt. The moisture content (MC) in the aggregate is calculated by Equation (2.5). Absorption is defined as the moisture content in the SSD condition.

$$MC = \frac{m_m - m_d}{m_d} \times 100\% \tag{2.5}$$

where MC = moisture content (%);

m_m = weight of aggregates with moisture (g);

m_d = weight of dry aggregates (g).

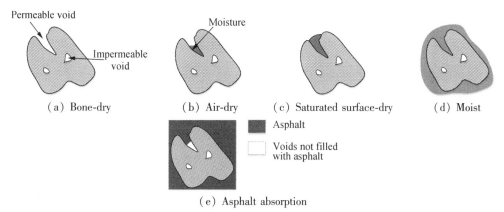

Figure 2.7 Voids and absorption conditions of the aggregate

2.2.5 Density

Aggregates have different density considering different volume and weight calculation methods. The mass-volume characteristics of aggregates are important for mix design. Density, the weight per unit volume, can be used for mix design. However, specific gravity, which is the weight of a material divided by the weight of an equal volume of water, is more commonly used.

Figure 2.8 is an aggregate particle coated with asphalt (grey color), including the aggregate solid, impermeable voids, and permeable voids. The permeable voids can be further divided into the portion filled with absorbed asphalt and the portion not filled with asphalt. We have three types of volumes. The apparent volume includes the volume of aggregate solid and impermeable voids, excluding the permeable voids. The bulk volume which is the volume at SSD status includes the volume of aggregate solid, impermeable voids, and permeable voids. The effective volume for aggregate in asphalt mixture includes the volume of aggregate plus the volume in the water permeable voids that are not filled with asphalt. We have two types of weight. The dry weight is the weight of the solid while the SSD weight includes both the weight of the solid and the weight of absorbed water in the permeable voids at SSD condition.

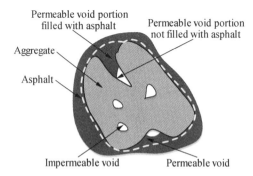

Figure 2.8 An aggregate particle coated with asphalt

Four types of specific gravity are defined based on how voids in the aggregate particles are considered. Apparent specific gravity is the weight of the aggregate over the apparent volume. Bulk specific gravity is the weight of the aggregate to the bulk volume. SSD specific gravity is the weight of the aggregates and water in the water-permeable voids to the bulk volume. Effective specific gravity is the weight of the aggregate to the effective volume. In this chapter, we are going to discuss the calculation of the first three density. The calculation of the effective specific gravity will be discussed in the chapter on asphalt mixture.

The specification JTG E42—2005 T0304 provides a method to measure the density and absorption of coarse aggregates. We immerse the samples in water for 24 hours, put them in the wire basket suspended in water, and measure the weight of aggregates underwater, denoted as m_w. It equals the weight of aggregates minus the weight of water with the same apparent volume. Then, we use a towel to remove the visible water film on the surface of aggregates and measure the SSD weight of aggregates, denoted as m_f, which equals the weight of aggregates plus the weight of water in the permeable voids.

After that, we dry the aggregates in the oven to a constant weight and measure the dry weight of aggregates, denoted as m_a.

1) Apparent density

The apparent specific gravity of coarse aggregates is calculated by Equation (2.6). It is the dry weight of coarse aggregates over the apparent volume. The apparent density of coarse aggregates can be calculated by Equation (2.7), which equals the density of water at testing temperature times the apparent specific gravity.

$$\gamma_a = \frac{m_a}{m_a - m_w} \tag{2.6}$$

$$\rho_a = \rho_w \gamma_a \tag{2.7}$$

where γ_a = apparent specific gravity of coarse aggregates;

m_a = dry weight of coarse aggregates (g);

m_w = weight of coarse aggregates underwater (g);

ρ_a = apparent density of coarse aggregates (g/cm^3);

ρ_w = density of water (g/cm^3).

2) Bulk density

The bulk specific gravity of coarse aggregates is calculated by Equation (2.8). It is the dry weight of coarse aggregates over the bulk volume. The bulk density of coarse aggregates can be calculated by Equation (2.9), which equals the density of water at testing temperature times the bulk specific gravity.

$$\gamma_b = \frac{m_a}{m_f - m_w} \tag{2.8}$$

$$\rho_b = \rho_w \gamma_b \tag{2.9}$$

where γ_b = bulk specific gravity of coarse aggregates;

m_f = SSD weight of coarse aggregates (g);

ρ_b = bulk density of coarse aggregates (g/cm^3).

3) SSD density

The SSD specific gravity of coarse aggregates is calculated by Equation (2.10). It is the SSD weight of coarse aggregates over the bulk volume. The SSD density of coarse aggregates can be calculated by Equation (2.11), which equals the density of water at testing temperature times the SSD specific gravity of coarse aggregates.

$$\gamma_s = \frac{m_f}{m_f - m_w} \tag{2.10}$$

Civil Engineering Materials

$$\rho_s = \rho_w \gamma_s \qquad (2.11)$$

where γ_s = SSD specific gravity of coarse aggregates;

ρ_s = SSD density of coarse aggregates (g/cm^3).

4) Absorption

Absorption can be calculated by Equation (2.12). It is the weight of water in the permeable voids over the dry weight of coarse aggregates.

$$w_x = \frac{m_f - m_a}{m_a} \times 100\% \qquad (2.12)$$

where w_x = absorption of coarse aggregates (%).

2.2.6 Unit Weight and Voids in Aggregates

Unit weight is used to estimate the weight of the aggregate by volume during purchase and transportation. According to specification JTG E42—2005 T0309, to measure the unit mass, we fill a rigid container of known volume with the aggregate and compact it by rodding, jigging, or shoveling, as shown in Figure 2.9. The unit weight is calculated by Equation (2.13). It is the weight of the aggregate over the volume of the container.

$$\rho_0 = \frac{m_0}{V_0} \qquad (2.13)$$

where ρ_0 = unit weight of coarse aggregates (g/cm^3);

m_0 = weight of coarse aggregates (g);

V_0 = volume of the container (cm^3).

The percentage of voids between aggregate particles can be calculated by Equation (2.14). It equals one minus the volume of the aggregate over the volume of the container, or one minus the ratio of unit weight over the density of the aggregate. According to specification JTG E42—2005, the bulk density is used to calculate the percentage of voids of asphalt mixture; while the apparent density is used to calculate the percentage of voids of Portland cement concrete.

$$Void\% = \begin{cases} \left(1 - \frac{V_0}{V_b}\right) \times 100\% = \left(1 - \frac{\rho_0}{\rho_b}\right) \times 100\%, & \text{asphalt mixture} \\ \left(1 - \frac{V_0}{V_a}\right) \times 100\% = \left(1 - \frac{\rho_0}{\rho_a}\right) \times 100\%, & \text{cement concrete} \end{cases} \qquad (2.14)$$

where $Void\%$ = percentage of voids (%);

V_b = bulk volume of coarse aggregates (cm^3);

V_a = apparent volume of coarse aggregates (cm^3);

ρ_b = bulk specific gravity of coarse aggregates (cm^3);

ρ_a = apparent specific gravity of coarse aggregates (cm^3).

(a) Unit weight test container　　(b) Aggregates packed in the container

Figure 2.9　Tests for unit weight and percentage of voids

When two or more aggregates from different sources are mixed, some of the properties of the mixed aggregates can be calculated from the properties of the individual component. For example, if we mix aggregates A and B at a specific ratio. The mixed bulk specific gravity is calculated based on the principle of equal volume, which means the volume of the mixed aggregates equals the sum of the volumes of aggregates A and B. The absorption of the mixed aggregates equals the sum of absorbed water of aggregates A and B.

Question:

Mix aggregates A and B at the ratio of 60:40, calculate (1) bulk specific gravity x and (2) absorption y of the mixed aggregates.

A: Bulk specific gravity = 2.952, absorption = 0.4%

B: Bulk specific gravity = 2.476, absorption = 5.2%

Answer:

(1) Calculate mixed bulk specific gravity x

$$\text{Mixed volume} = \text{Volume of A} + \text{Volume of B}$$

$$100/x = 60/2.952 + 40/2.476$$

$$x = 2.741$$

(2) Calculate mixed absorption y

$$\text{Mixed water content} = \text{Water content of A} + \text{Water content of B}$$

$$100y = 60 \times 0.4 + 40 \times 5.2$$

$$y = 2.32$$

2.2.7 Angularity and Flakiness

The shape and surface texture of the individual aggregate particle determines the packing density and the mobility of the aggregate in a mixture. There are two characterizations of the shape of the aggregate particle: angularity and flakiness. As shown in Figure 2.10, coarse aggregates can be classified into angular and rounded based on their angularity. Angular and rough-textured aggregates produce bulk materials with higher stability than rounded and smooth-textured aggregates. However, the angular aggregates are more difficult to work into place than rounded aggregates, since their shapes make it difficult for them to slide across each other. Based on the flakiness, coarse aggregates can be classified as flaky, elongated, and flaky and elongated. The content of flaky and elongated particles should be limited in a good quality aggregate since they tend to break under loads and prevent the development of a strong aggregate skeleton (Figure 2.11).

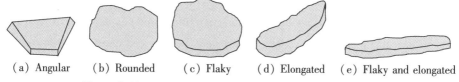

(a) Angular   (b) Rounded   (c) Flaky   (d) Elongated   (e) Flaky and elongated

Figure 2.10　Angularity and flakiness of coarse aggregates

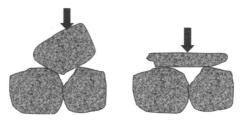

Figure 2.11　Flaky and elongated particles tending to break under loads

In specification ASTM D4791-19, a particle is defined as "flat" if the ratio of the middle dimension to the smallest dimension exceeds 3 to 1. A particle is defined as "elongated" if the ratio of the longest dimension to the middle dimension exceeds 3 to 1. In the Superpave criteria, particles are classified as "flat and elongated" if the ratio of the largest dimension to the smallest dimension exceeds 5 to 1. The specification JTG E42—2005 T0311 provides a method to determine the flaky and elongated particles in coarse aggregate using two apparatus shown in Figure 2.12.

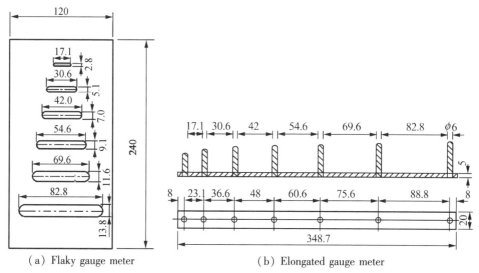

(a) Flaky gauge meter (b) Elongated gauge meter

Figure 2.12 Flaky and elongated particle gauge meter (unit: mm)

Table 2.6 shows a specific amount of samples required for the test, depending on the nominal maximum size. The specification JTG E42—2005 T0312 introduces a method by measuring each aggregate particle with a caliper. A particle is defined as flaky or elongated if the ratio of the longest dimension to the smallest dimension exceeds 3 to 1. The weight percentage of flaky and elongated particles is usually calculated to evaluate the quality of the aggregate.

$$Q_e = \frac{m_1}{m_0} \times 100\% \tag{2.15}$$

where Q_e = content of flaky and elongated particles (%);
m_1 = weight of flaky and elongated particles (g);
m_0 = weight of the sample (g).

Table 2.6 The weight of samples required for the flaky and elongated particle test (JTG E42—2005 T0311)

Nominal maximum size (mm)	9.5	16	19	26.5	31.5	37.5	53	63
Minimum weight of samples (kg)	0.3	1	2	3	5	10	10	10

2.2.8 Fractured Faces and Crushed Particle

Fractured faces are angular, rough, or broken surfaces of an aggregate particle created by crushing, by other artificial means, or by nature. As shown in Figure 2.13, a fractured face should be no less than 25% of the maximum cross-sectional area of the crushed particle. The number of fractured faces is directly related to the shape and surface texture of coarse aggregates. A crushed particle is defined as a particle of the aggregate having at least one fractured face. The specification JTG E42—2005 T0346 introduces a method to test the percentage of crushed particles of coarse aggregates. For a specific amount of samples, the fractured faces of each particle need to be checked, as shown in Table 2.7. Then the percentage of crushed particles can be calculated by Equation (2.16). It is noted that Q should be less than 15%.

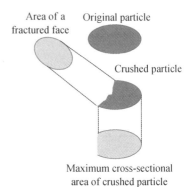

Figure 2.13 Aggregates with a fractured face

$$P = \frac{F + Q/2}{F + Q + N} \times 100\% \quad (2.16)$$

where P = percentage of crushed particles(%);

F = weight of the particles with at least one fractured face(g);

Q = weight of the particles that are difficult to determine if they include fractured faces(g);

N = weight of the particles without a fractured face(g).

Table 2.7 The weight of samples required for the fractured faces test

Nominal maximum size (mm)	9.5	13.2	16	19	26.5	31.5	37.5	50
Minimum weight of samples (kg)	0.2	0.5	1	1.5	3	5	7.5	15

2.3 Mechanical Properties of Coarse Aggregates

The mechanical properties of coarse aggregates include crushing value, impact value, polished stone value, abrasion value, soundness, and the alkali reaction.

2.3.1 Crushing Value

The crushing value measures the strength of the aggregates under gradually applied compressive loads when they are used in pavements. It is the weight percentage of the crushed materials obtained when the test aggregates are subjected to a specified load. According to specification JTG E42—2005 T0316, 3 kg of coarse aggregates with sizes between 9.5 and 13.2 mm are air- or oven-dried and then put in the container for the test. A compressive load is applied at a uniform rate till 400 kN in 10 minutes and then hold for 5 seconds (Figure 2.14). The passing percentage at the 2.36 mm size of the crushed aggregates is calculated. Generally, aggregates with a crushing value less than 10% are strong aggregates while those with a crush value higher than 35% are weak aggregates.

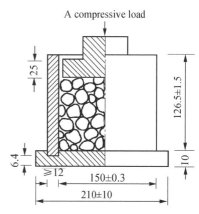

Figure 2.14 Crushing value tester (unit: mm)

2.3.2 Impact Value

Impact value measures the resistance of aggregates to sudden shock or impact especially when they are used in pavements. In specification JTG E42—2005 T0322, coarse aggregates with sizes between 9.5 and 13.2 mm are placed in the cup in three layers and each layer is compacted by 25 strokes of the tamping rod, as shown in Figure 2.15. Then, the hammer is raised 380 mm above the upper surface of the aggregates in the cup and falls freely onto the aggregates. The passing percentage at the 2.36 mm size of the crushed aggregate is calculated as the impact value.

2.3.3 Polished Stone Value

Polished stone value (PSV) measures the resistance of coarse aggregates to the polishing of vehicle tires for

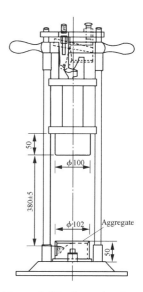

Figure 2.15 Impact value tester (unit: mm)

pavement surface course. High PSV indicates good resistance to the polishing of vehicle tires. In specification JTG E42—2005 T0321, the PSV test involves two tests: the polishing test and the skid resistance test. The coarse aggregates are firstly polished by an accelerated polishing machine as shown in Figure 2.16. This machine polishes samples of aggregates, simulating actual road conditions, and is used in conjunction with the portable skid resistance tester to determine the PSV. After polishing, the PSV is tested by the pendulum skid tester. During the test, the pendulum arm is raised and latched onto a rigid arm on the tester and then released. The pendulum swings down and drags the spring-loaded rubber block over the convex face of the sample. The needle indicator stops and remains at the highest point of the swing, which is recorded as the PSV. Table 2.8 presents the requirements for the PSV of aggregates used in the pavement surface course. The aggregate should have higher PSV when they are used in high-grade roads and humid areas. Table 2.9 shows the typical PSV of rocks, mostly meeting the requirements in Table 2.8. In China, basalt is more widely used in the surface course of expressways.

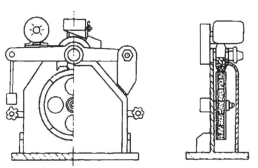

Figure 2.16 Polished stone value tester

Table 2.8 Requirements of PSV for aggregates used in pavements

Annual rainfall (mm)	Expressway and Class I road	Other roads
>1000	>42	>40
500 – 1000	>40	>38
250 – 500	>38	>36
<250	>36	...

Table 2.9 Typical PSV of rocks

PSV	Limestone	Hornfels	Porphyry	Quartzite	Granite	Basalt
Average	43	45	56	58	59	62
Range	30 – 70	40 – 50	43 – 71	45 – 67	45 – 70	45 – 81

2.3.4 Abrasion Value

Abrasion value measures the resistance of aggregates to surface wear by abrasion. There are two types of abrasion tests. In specification JTG E42—2005 T0323, the Dorry abrasion tester is used to measure a cylindrical specimen with a height of 25 cm and a diameter of 25 cm subjected to the abrasion against a rotating metal disk sprinkled with quartz sand, as shown in Figure 2.17. The weight loss of the cylinder after 1000 revolutions of the table is determined as the Dorry abrasion. In specification JTG E42—2005 T0317, the Los Angeles (LA) abrasion tester is used to measure a coarse aggregate sample subjected to abrasion, impact, and grinding in a rotating steel drum containing a specified number of steel spheres (Figure 2.18). Different sizes and weights of samples are tested depending on the maximum aggregate size. The passing percentage at 1.7 mm of the crushed aggregate is calculated as the LA abrasion value. The LA abrasion test is more widely used.

Figure 2.17 Samples of the Dorry abrasion test   **Figure 2.18** LA abrasion tester

2.3.5 Soundness

Soundness is the ability of aggregates to withstand weathering. The soundness test simulates weathering by soaking aggregates in a saturated sodium sulfate solution. These sulfates form crystals which grow in the aggregates when dried, simulating the effect of freezing. According to specification JTG E42—2005 T0314, a specific number of aggregate samples is firstly obtained, depending on the nominal maximum size of the aggregate as shown in Table 2.10. The test involves subjecting aggregates to 5 cycles of soaking in the sulfate for 16 hours, followed by drying. Then, the average weight loss is calculated to evaluate the soundness. Figure 2.19 shows the cracking caused by the expansion of sodium sulfate crystals. During the soaking process, the Na_2SO_4 solution infiltrates into permeable voids. Then, during the drying process, the Na_2SO_4 solution forms crystals, causing expansion.

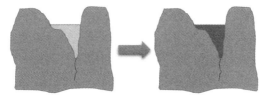

Figure 2.19 Soundness test of aggregates

Table 2.10 The weight of sample size required for the soundness test

Nominal maximum size (mm)	2.36 – 4.75	4.75 – 9.5	9.5 – 19	19 – 37.5	37.5 – 63	63 – 75
Sample weight (kg)	0.5	0.5	1	1.5	3	5

2.3.6 Alkali Reaction

Alkali reaction is a swelling reaction in concrete between the highly alkaline cement paste and the reactive silica in aggregates in a moist environment. The soluble and viscous sodium silicate hydrate (SSH) gel formed during the reaction can absorb water and increase in volume, causing cracking of the concrete, as shown in Figure 2.20. The reactivity of alkali-reactive aggregates can be minimized by limiting the alkali content of the cement. The reaction can also be reduced by keeping the concrete structure as dry as possible. Fly ash, ground granulated blast furnace slag, silica fume, or natural pozzolans can be used to reduce the alkali reaction.

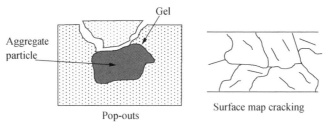

Figure 2.20 Alkali reaction

The specification JTG E42—2005 T0347 introduces a method to calculate the relative content of hydrogen ions in aggregates to evaluate the alkali reaction of aggregates. The specification JTG E42—2005 T0325 introduces a method to test the expansion of a 25.4 mm × 25.4 mm × 285 mm cement mortar beam at different curing times to evaluate the alkali-reactive potential of the aggregate.

2.4 Properties of Fine Aggregates

Fine aggregates have some common properties as coarse aggregates including density, angularity, and unit weight, although the test methods are different. In addition, fine aggregates have some specific properties such as fineness modulus, sand equivalency, and deleterious materials test.

2.4.1 Density

Same as coarse aggregates, the apparent, bulk, SSD, and effective specific gravity or density are also very important for fine aggregates. The mass-volume characteristics of coarse aggregates also apply to fine aggregates.

To test the density and absorption of fine aggregates, according to JTG E42—2005 T0328, we first measure the weight of the pycnometer filled with water to the calibration mark and record it as m_1. It includes the weight of the pycnometer, water of the same apparent volume of fine aggregates, and water of the rest volume in the pycnometer. We soak a representative sample of fine aggregates in water for 24 hours, dry it back to the SSD condition, weigh around 500 g sample, and record it as m_3, which includes the weight of aggregates and water in the permeable voids. Then, we place the SSD sample in a pycnometer and add water to the constant volume mark on the pycnometer and the weight is determined again as m_2, which includes the weight of the pycnometer, fine aggregates, and water of rest volume in the pycnometer. The sample is dried, and the weight is determined and recorded as m_0.

For coarse aggregates, we use a towel to dry the aggregates to a saturated surface dry condition. For fine aggregates, using the towel to dry the small particles is difficult and we can use the hair drier instead. As shown in Figure 2.21, to determine if the sample is at the SSD condition, we put the partially dried fine aggregates into the mold and tamp them. Then we lift the mold vertically, if the surface moisture still exists, the fine aggregates will retain the molded shape. If this is the case, allow the sand to dry and repeat checking until the fine aggregates slump slightly, indicating that it has reached a surface-dry condition.

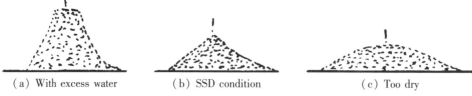

(a) With excess water (b) SSD condition (c) Too dry

Figure 2.21 Determination of SSD condition of fine aggregates

1) Apparent density

The apparent specific gravity of fine aggregates is calculated by Equation (2.17). $(m_0 + m_1 - m_2)$ is the weight of the water of the same apparent volume of fine aggregates which equals the apparent volume of fine aggregates. The apparent density of fine aggregates can be calculated by Equation (2.18), which equals the density of water at testing temperature times the apparent specific gravity of fine aggregates.

$$\gamma_a = \frac{m_0}{m_0 + m_1 - m_2} \tag{2.17}$$

$$\rho_a = \rho_w \gamma_a \tag{2.18}$$

where γ_a = apparent specific gravity of fine aggregates;

m_0 = dry weight (g/cm^3);

m_1 = weight of pycnometer filled with water (g);

m_2 = weight of the pycnometer filled with aggregates and water (g);

ρ_a = apparent density of fine aggregates (g/cm^3);

ρ_w = density of water (g/cm^3).

2) Bulk density

The bulk specific gravity of fine aggregates is calculated by Equation (2.19). $(m_3 + m_1 - m_2)$ is the weight of the water of apparent volume of fine aggregates and the weight of the water in permeable voids, which equal the bulk volume of fine aggregates. The bulk density of fine aggregates can be calculated by Equation (2.20), which equals the density of water at testing temperature times the bulk specific gravity of fine aggregates.

$$\gamma_b = \frac{m_0}{m_3 + m_1 - m_2} \tag{2.19}$$

$$\rho_b = \rho_w \gamma_b \tag{2.20}$$

where γ_b = bulk specific gravity of fine aggregates;

m_3 = SSD weight of fine aggregates (g);

ρ_b = bulk density of fine aggregates (g/cm^3).

3) SSD density

The SSD specific gravity of fine aggregates is calculated by Equation (2.21). The SSD density of fine aggregates can be calculated by Equation (2.22), which equals the density of water at testing temperature times the SSD specific gravity of fine aggregates.

$$\gamma_s = \frac{m_3}{m_3 + m_1 - m_2} \tag{2.21}$$

$$\rho_s = \rho_w \gamma_s \tag{2.22}$$

where γ_s = SSD specific gravity of fine aggregates (g);

ρ_s = SSD density of fine aggregates (g/cm^3).

4) Absorption

Absorption can be calculated by Equation (2.23). It is the weight of water in the permeable voids over the dry weight of fine aggregates.

$$w_x = \frac{m_3 - m_0}{m_0} \times 100\% \tag{2.23}$$

where w_x = absorption of fine aggregates (%).

2.4.2 Unit Weight and Void Ratio

According to specification JTG E42—2005 T0331, the unit weight of fine aggregates can be tested with a standard funnel and container (as shown in Figure 2.22) and calculated by Equation (2.24)

$$\rho_0 = \frac{m_0}{V_0} \tag{2.24}$$

where ρ_0 = unit weight of the fine aggregates (g/cm^3);

m_0 = weight of fine aggregates (g);

V_0 = volume of the container (cm^3).

The percentage of voids between aggregate particles, $Void\%$, can be calculated by one minus the volume of aggregates over the volume of the container, which equals one minus the ratio of unit weight over the apparent density.

$$Void\% = \left(1 - \frac{\rho_0}{\rho_a}\right) \times 100\% \tag{2.25}$$

where ρ_a = apparent density of the fine aggregates (g/cm^3).

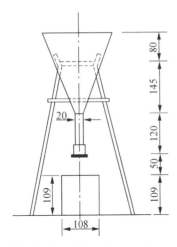

Figure 2.22 Unit weight tester for fine aggregates (unit: mm)

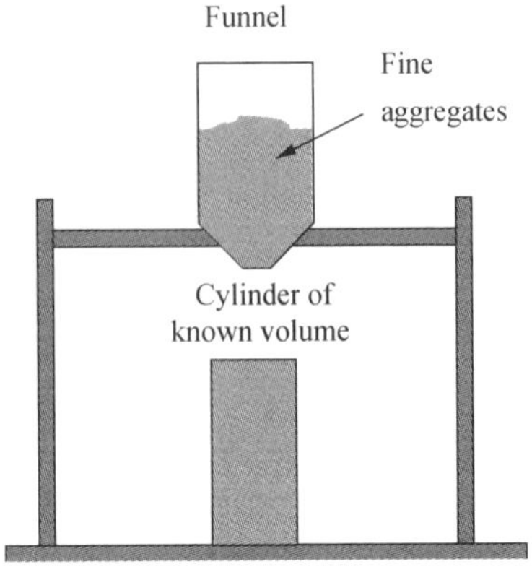

Figure 2.23 Angularity test of fineness

2.4.3 Angularity

Fine aggregates' angularity is important because an excess of rounded fine aggregates may cause asphalt pavement rutting. According to specification JTG E42—2005 T0344, the test estimates angularity by measuring the loose uncompacted void content of a fine aggregate sample using a funnel as shown in Figure 2.23. The loose uncompacted void content is indicative of the relative angularity and surface texture of the sample. The higher the void content, the higher the assumed angularity and rougher the surface.

2.4.4 Fineness Modulus

The fineness modulus is a measure of the fine aggregates' gradation and is a critical factor for the mix design of Portland cement concrete. The fineness modulus for fine aggregates used in cement concrete is calculated by Equation (2.26). Fine aggregates with a fineness modulus between 3.1 and 3.7 are regarded as coarse sands, fine aggregates with a fineness modulus between 2.3 and 3.0 are regarded as medium sand, while fine aggregates with a fineness modulus between 1.6 and 2.2 are regarded as fine sand.

$$M_X = \frac{A_{0.15} + A_{0.3} + A_{0.6} + A_{1.18} + A_{2.36} - 5A_{4.75}}{100 - A_{4.75}} \quad (2.26)$$

where M_X = fineness modulus;

$A_{0.15}, A_{0.3}, A_{0.6}, A_{1.18}, A_{2.36}, A_{4.75}$ = cumulative percentage retained on sieve with size of 0.15, 0.3, 0.6, 1.18, 2.36, and 4.75 mm.

As shown in Equation (2.27), the fineness modulus for asphalt mixtures and pavement base is an empirical figure obtained by adding the cumulative percentage retained on each of a specified series of sieves, and dividing the sum by 100. Therefore, a high fineness modulus means more aggregates are retained on those sieves and therefore the size of aggregates is larger. Fine aggregates with a fineness modulus between 2.9 and 3.2 are regarded as coarse sand, a fineness modulus between 2.6 and 2.8 are regarded as medium sand, while a fineness modulus between 2.2 and 2.5 are regarded as fine sand.

$$M_X = \frac{A_{0.15} + A_{0.3} + A_{0.6} + A_{1.18} + A_{2.36} + A_{4.75}}{100} \tag{2.27}$$

2.4.5 Sand Equivalency

Sand equivalency (SE) is the relative volume ratio of sand over all particles in fine aggregates. It indicates the relative volume of sand in fine aggregates. As shown in Figure 2.24, according to specification JTG E42—2005 T0334, a small amount of flocculating solution is poured into a graduated cylinder and is agitated to loosen the clay-like coatings from the sand particles. The sample is then irrigated with additional flocculating solution forcing the clay-like material into suspension above the sand. The sand equivalency is expressed as a ratio of the height of sand over the height of the clay top.

$$SE = \frac{h_{sand}}{h_{clay}} \times 100 \tag{2.28}$$

where h_{clay} = height of the clay (mm);
h_{sand} = height of the sand (mm).

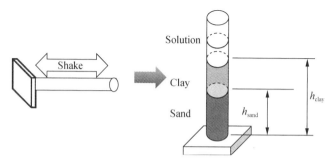

Figure 2.24 Sand equivalency test

2.4.6 Methylene Blue Test

The methylene blue test of fine aggregates is a measure of the amount of potentially harmful fine materials such as clay and organic materials. According to specification JTG E42—2005 T0349, the methylene blue value is a function of the amount and characteristics of clay minerals. High methylene blue values indicate increased potential for diminished fine aggregates or mineral filler performance in a cementitious mixture due to the presence of clay.

2.5 Filler

Fillers are defined as aggregate particles smaller than 0.075 mm sieve. In asphalt mixtures, fillers are mixed with asphalt binder to form asphalt mortar, which can improve the bond between aggregates and asphalt, and increase the stability or rutting resistance of asphalt mixtures. It can also fill the voids in the asphalt mixtures and thus reduce the asphalt required. Table 2.11 summarizes the requirements of mineral fillers used in asphalt mixtures and the corresponding test methods.

Table 2.11 Requirements for mineral fillers for asphalt mixtures

Properties	Expressway and first-class road	Other roads	Test methods
Min. apparent specific gravity	2.50	2.45	T 0352
Max. absorption (%)	1	1	T 0103
Proportion <0.6mm (%)	100	100	
Proportion <0.15mm (%)	90 – 100	90 – 100	T 0351
Proportion <0.075mm (%)	75 – 100	70 – 100	
Appearance	No agglomeration	—	—
Max. hydrophilic coefficient	1	—	T 0353
Max. plastic index (%)	4	—	T 0354
Thermal stability	Normal	—	T 0355

2.6 Gradation Design

Gradation describes the particle size distribution which is an important attribute of aggregates. Gradation design is to determine the optimal particle size distribution of aggregates when used in Portland cement concrete and asphalt mixtures, or as an unbound material.

2.6.1 Gradation Curves

A typical gradation curve shows the passing percentage of aggregates at different sieve sizes. In a gradation curve chart, the horizontal axis is the sieve size usually in a logarithm scale or 0.45 power scale. The effects of log or 0.45 power transform can spread out the small values and bring large values closer together. The vertical axis is the passing percentage. Figure 2.25 uses a logarithm horizontal axis to expand the small numbers so that the small size sieves can be differentiated in the axis.

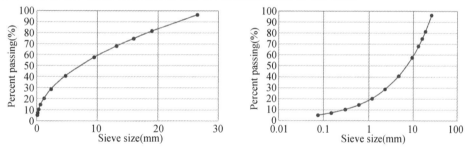

Figure 2.25 The normal horizontal axis (left) and the logarithm horizontal axis (right)

Figure 2.26 shows different aggregates or gradations identified in the gradation chart. The coarse aggregate is on the right side of the chart while the fine aggregate is on the left side of the chart. The mineral filler is also on the left part of the chart, but narrower than the region of the fine aggregate. Mineral dust is even smaller and is on the left narrower part of the chart. Fine gradation lies above the diagonal line or the 0.45 power curve because it has a higher passing percentage at most sieves. Coarse gradation lies below the 0.45 power curve because it has a lower passing percentage at most sieves.

Figure 2.27 shows the typical gradation curves of aggregates and Figure 2.28 shows the cross-sectional view of the mixes. In addition to the traditional dense gradation curve, aggregates can have other characteristic distributions. A uniform or one-sized

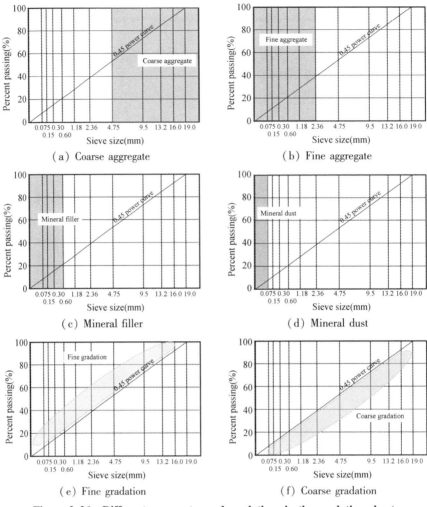

Figure 2.26 Different aggregates and gradations in the gradation charts

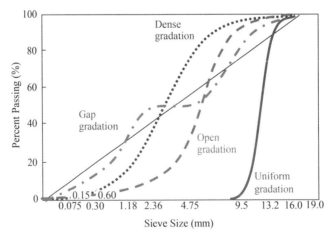

Figure 2.27 Three gradation curves

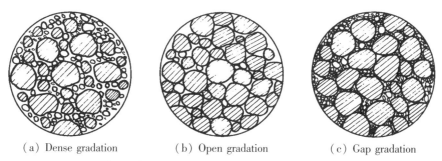

(a) Dense gradation (b) Open gradation (c) Gap gradation

Figure 2.28 **Aggregates of different gradations**

distribution has the majority of aggregates passing one sieve and being retained on the next smaller sieve. Hence, the majority of the aggregates have essentially the same diameter and their gradation curve is nearly vertical. One-sized graded aggregates will have good permeability, but poor stability, and are used in such applications as chip seals of pavements. Open-graded aggregates are missing small aggregate sizes that would block the voids between the larger aggregates. Since there are a lot of voids, the material will be highly permeable, but may not have good stability. Gap-graded aggregates are missing one or more sizes of material. Their gradation curve has a near-horizontal section, indicating that nearly the same portions of the aggregates pass two sieves of different sizes.

Figure 2.29 shows the dense and open gradation asphalt mixtures on rainy days. The left-side pavement uses dense gradation; the right-side pavement uses open gradation.

(a) Dense gradation (b) Open gradation

(c) The dense (left) and open (right) gradation asphalt mixtures

Figure 2.29 **Pavement surface with dense and open gradation asphalt mixtures**

Civil Engineering Materials

The open gradation asphalt mix has much less water film and splashing, and therefore the riding safety on rainy days is greatly improved. Some research found that the open gradation asphalt mix can significantly reduce the traffic accident rate in a long term.

2.6.2 Gradation Theory

There have been several theories to help design aggregate gradation including maximum density theory, the Barley method, and the particle intervention theory. The maximum density theory is the most widely used, in which the density of an aggregate mix is a function of the size distribution of the aggregates. In 1907, Fuller and Thompson established the relationship to determine the distribution of aggregates that provides the maximum density or minimum amount of voids.

$$p^2 = kd \tag{2.29}$$

where p = passing percentage (%);
k = coefficient;
d = sieve size (mm).

Later, Fuller improved the maximum density curve equation (2.30) with the power function. The value of the exponent n recommended by Fuller is 0.5. In the 1960s, the US Federal Highway Administration (FHWA) recommended a value of 0.45 for n and introduced the "0.45 power" gradation chart in the Strategic Highway Research Program (SHRP) for asphalt mixture design. Portland cement concrete usually uses a value of 0.35 for n, as shown in Figure 2.30.

$$p = 100\left(\frac{d}{D}\right)^n \tag{2.30}$$

where n = power number;
D = maximum size of the aggregate (mm).

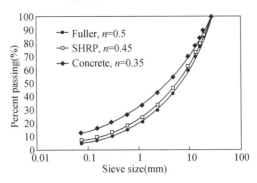

Figure 2.30 The maximum density curves

2.6.3 Gradation Design

A single aggregate source is unlikely to meet the gradation requirements. One way for gradation design is to prepare the aggregates at each single sieve size and mix them at the designed proportion. This means that for n sieves, n groups of aggregates of the sieve sizes need to be prepared, which is costly. Therefore, the blending of aggregates from two to five sources is commonly used. As shown in Table 2.12, the objective of gradation design is to determine the optimum proportions of the four aggregates to let the mixed gradation fall within the required range and as close as possible to the designed gradation which is the mid-value of the lower and upper limits. The proportion can be determined by the graphical method and the numerical method.

Table 2.12 Gradation design template

Sieve size (mm)	4#	3#	2#	1#	Blended	Designed	Lower limit	Upper limit
37.5	100	100	100	100	100	100	100	100
25	100	100	100	100	100	100	100	100
19	96	100	100	100	99.2	98	96	100
12.5	88	100	100	100	97.6	97.5	95	100
9.5	49	87	100	100	87.85	86	82	90
4.75	9	56	100	100	75.2	75	70	80
2.36	2	13	90	98	64.05	65	60	70
1.18	2	6	57	46	33.95	35	30	40
0.6	2	3	36	31	22.25	23	20	26
0.3	2	2	24	22	15.5	15	10	20
0.15	2	2	12	13	8.9	9	6	12
0.075	2	1	6	5	4.05	4	2	6
Proportion (%)	20	15	25	4	—	—	—	—

1) Graphical method

As shown in Figure 2.31(a), we have four aggregates with different sizes and their gradation curves have neither overlappings nor spaces. The diagonal line is the objective gradation. In the mixed aggregates, those larger than size A are all from aggregate A. We can draw a vertical line at sieve size A and a horizontal line starting from the crossover point of the vertical line and the objective gradation curve. We can determine that the proportion of aggregate A is X. Then, the portion larger than size B in the mixed aggregates are all from aggregates A and B and the proportion of aggregate A is already known, therefore the rest is the proportion of aggregate B. Similarly, we can find that the proportion of aggregate B is Y, and the proportion of aggregate C is Z. The rest proportion W is for aggregate D. Usually, there are overlappings or spaces between the gradation curves of aggregates. As shown in Figure 2.31(b), we can use the mid-point between the two curves to estimate the corresponding proportions.

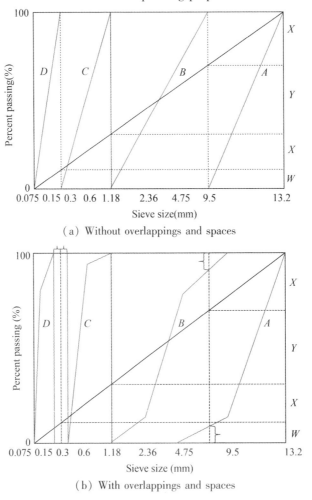

Figure 2.31 Graphical method for gradation design

2) Numerical method

The numerical method uses the trial and error procedure to determine the proportions.

For k types of aggregates, at sieve No. 1, the sum of passing percentage of the k aggregates, $p_{1(1)} \cdot x_1 + p_{2(1)} \cdot x_2 + \cdots + p_{k(1)} \cdot x_k$ should be equal to the designed passing percentage $p_{(1)}$ at sieve 1. Similarly, at sieve No. n, the sum of passing percentage of the k aggregates, $p_{1(n)} \cdot x_1 + p_{2(n)} \cdot x_2 + \cdots + p_{k(n)} \cdot x_k$ should be equal to the designed passing percentage $p_{(n)}$. For a total of n sieves, we have a total of n equations (2.31). The trial and error procedure or the programming optimization method can be adopted to find the optimal ratio of x_1, x_2 and x_k to determine the gradation.

$$\begin{aligned} p_{1(1)} \cdot x_1 + p_{2(1)} \cdot x_2 + \cdots + p_{k(1)} \cdot x_k &= p_{(1)} \\ p_{1(2)} \cdot x_1 + p_{2(2)} \cdot x_2 + \cdots + p_{k(2)} \cdot x_k &= p_{(2)} \\ &\cdots \\ p_{1(n)} \cdot x_1 + p_{2(n)} \cdot x_2 + \cdots + p_{k(n)} \cdot x_k &= p_{(n)} \end{aligned} \quad (2.31)$$

Questions

1. Discuss the three main types of crushers to produce aggregates and which one is preferred when producing small-size aggregates.

2. Define the fineness modulus of aggregates and discuss how to classify the sand based on its fineness modulus.

3. Discuss how to measure the angularity of coarse and fine aggregates.

4. Discuss how to measure the crushing value, impact value, polished stone value, and abrasion value.

5. Discuss the difference in calculating voids ratios for coarse aggregates used in Portland cement concrete and asphalt mixture and explain why.

6. For the two samples of fine aggregates shown below, calculate the moisture content for each sample, and the moisture content of the mix if they are mixed at a ratio of 1:2.

Table 2.13 Measured weights of two samples of fine aggregates

Measures	Samples	
	A	B
Wet weight (g)	520.1	521.6
Dry weight (g)	490.5	491.3
Absorption (%)	2.6	2.7

7. For a sample of coarse aggregates from a stockpile, the following weights are found:

Weight of moist aggregate sample as brought to the laboratory = 5289 g

Weight of oven-dried aggregates = 5205 g

Weight of aggregates submerged in water = 3288 g

Weight of SSD aggregates = 5216 g

Calculate

(1) The bulk specific gravity

(2) The apparent specific gravity

(3) The moisture content of stockpile aggregate

(4) Absorption

8. For a sample of fine aggregates from a stockpile, the following weights are found:

Weight of SSD sand = 500.0 g

Weight of pycnometer with water only = 621.7 g

Weight of pycnometer with sand and water = 935.2 g

Weight of dry sand = 96.1 g

Calculate

(1) The bulk specific gravity

(2) The apparent specific gravity

(3) The SSD specific gravity

(4) Absorption

9. For two coarse aggregates which have to be blended. The results are as follows:

Aggregate A: Bulk specific gravity = 2.796; absorption = 0.5%

Aggregate B: Bulk specific gravity = 2.468; absorption = 5.1%

Calculate

(1) What is the specific gravity of a mixture of 50% aggregate A and 50% aggregate B by weight?

(2) What is the absorption of the mixture?

10. Make a spreadsheet blend template in Excel to perform a gradation analysis with the five aggregates including fillers for the following lower and upper limits.

Table 2.14 Gradations of the five aggregates and design limits

Sieve size (mm)	Passing (%)							
	4#	3#	2#	1#	Filler	Mix	Lower limit	Upper limit
26.5	83.5	100	100	100	100	—	95	100
19	30.4	100	100	100	100	—	75	90
16	15.1	99.3	100	100	100	—	62	80
13.2	1.6	70	100	100	100	—	53	73
9.5	0	53.3	100	100	100	—	43	63
4.75	0	15	80	100	100	—	32	52
2.36	0	0	30	98.3	100	—	25	42
1.18	0	0	5	60	100	—	18	32
0.6	0	0	0	40	100	—	13	25
0.3	0	0	0	15	98.6	—	8	18
0.15	0	0	0	10	93.3	—	5	13
0.075	0	0	0	0.1	81.9	—	3	7

3 Inorganic Binding Materials

Binding materials can be classified into organic and inorganic materials according to their chemical composition. Organic binding materials in civil engineering mainly include asphalt and resin. Inorganic binding or cementing materials in civil engineering are mainly inorganic powder materials that can be mixed with water or the aqueous solution to form the slurry and gradually get hardened to form the artificial stone with strength after a series of physical and chemical changes. Inorganic binding materials can be classified into water-hardening and air-hardening binding materials according to their hardening conditions. Lime and gypsum are air-hardening materials which only harden in the air and cannot harden in the water; whereas Portland cement and aluminate cement, etc. are water-hardening materials that can harden in both air and water. This chapter covers the production, slaking, hardening, and properties of lime; and the production, hydration, properties, corrosion, and supplementary materials of cement.

3.1 Lime

3.1.1 Production

Lime has been used in building construction since at least 10,000 years ago because of its wide range of raw materials, low cost, and simple process. Egyptians used mortar made with a binder obtained by gypsum calcination in the pyramids of Giza more than 4000 years ago. Chinese started using lime in building foundations and roofs around 3000 years ago, and the mix of lime, clay, and sand slurry more than 2000 years ago. In 1812, the French established hydraulic lime manufacture from synthetic mixtures of limestone and clay.

3 Inorganic Binding Materials

Lime is produced from limestone, dolomite, or other natural materials mainly containing calcium and/or magnesium carbonate. By calcining the above materials, the calcium carbonate breaks down into calcium oxide and forms the quicklime. The quicklime is in block shape, also known as block ash. Its structure is crisp and is ground into quicklime powder for construction. This process is shown in Equation (3.1). The calcination temperature is often controlled at 900 – 1000 °C to accelerate the decomposition of limestone with varying density, size, and impurity.

$$CaCO_3 \xrightarrow{900 - 1000 \ °C} CaO + CO_2 \qquad (3.1)$$

The quicklime contains a specific amount of magnesium oxide from the magnesium carbonate in the limestone or the dolomite. The quicklime with less than 5% magnesium oxide is calcitic lime, while the quicklime with more than 5% magnesium oxide is dolomite lime. Compared with calcitic lime, dolomite lime is slower to mature but slightly stronger after hardening.

The quicklime may contain under-burnt lime and over-burnt lime. The under-burnt lime contains a specific amount of raw calcium carbonate or limestone residual in the lime due to the low calcination temperature. It generates less mortar when mixed with water and the quality is poor. The over-burnt lime contains over calcined particles which are less hydraulically reactive. Because of the high calcination temperature or longer calcination time, the surface of over-burnt lime particles is melted, forming a relatively dense structure and covering the lime particles with a layer of brown melt. The over-calcined particles harden slower than lime and may cause volume expansion, leading to uplifting and cracking of the hardened lime paste. Figure 3.1 shows the specific surface area of lime at different calcination temperatures and times.

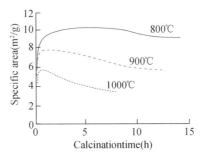

Figure 3.1 The specific surface area of lime at different calcination temperatures and times

3.1.2 Slaking and Hardening

To eliminate the damage caused by the over-burnt lime, lime putty is produced by slaking quicklime with an excess of water for several weeks until a creamy texture is produced. Lime releases a lot of heat when slaking which is nine times that of Portland cement, and the volume increases 1 - 2.5 times. The slaking process is the reaction of calcium and water to form calcium hydroxide, as shown in Equation (3.2). Slaked lime has to be immersed in water to avoid being carbonized.

$$CaO + H_2O \longrightarrow Ca(OH)_2 + Q \qquad (3.2)$$

Hardening of lime slurry includes crystallization and carbonization. The crystallization is when lime mortar dries, the calcium hydroxide solution is over-saturated and precipitates crystals. This process contributes very low strength. In addition, during the drying process of lime slurry, the pore structure is formed due to the evaporation of water. Due to the surface tension, the free water left in the pore forms a concave meniscus at the narrowest part of the pore, generating capillary pressure and making the lime particles more compacted and obtaining strength. This strength is similar to the strength gained by clay after water loss but it may lose this strength when it encounters water again.

Carbonation is the process that lime reacts with carbon dioxide and water, which gradually reconverts calcium oxide back to calcium carbonate, as shown in Equation (3.3). If the water content is too low, the carbonation process will stop. If the water content is too high and the pores are filled with water, the carbonation will only occur on the surface where the lime can contact carbon dioxide. When the calcium carbonate formed on the surface reaches a specific thickness, it prevents the infiltration of water and carbon dioxide, and the carbonation becomes extremely slow. Figure 3.2 shows the life cycle of lime, in which limestone (calcium carbonate) is converted to quicklime by heating, then to slaked lime by hydration, and naturally reverts to calcium carbonate by carbonation.

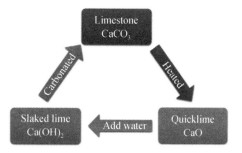

Figure 3.2 The life cycle of lime

$$Ca(OH)_2 + CO_2 + nH_2O \rightleftharpoons CaCO_3 + (n+1)H_2O \qquad (3.3)$$

3.1.3 Properties

Lime mortar or cement lime mortar prepared with limestone plaster or hydrated lime powder are among the most commonly used materials in buildings. Lime, cement, and slag stabilized mixtures are used as pavement base or subbase. Lime stabilized soil and lime piles are used in the foundation. Lime has the following properties, making it a suitable material for many applications in civil engineering.

(1) Good plasticity and water retention: Due to a large number of fine calcium hydroxide particles and the large specific surface area, a large amount of water can be absorbed on the surface of particles, making the mortar with good plasticity and water retention.

(2) Slow setting and low strength: Because of the low content of carbon dioxide in the air and the insulation of calcium carbonate formed on the surface, the formation of strength is very slow. The hardened lime has low strength, and its 28-day strength is 0.2 - 0.5 MPa. Lime mortar can take a long time to achieve its full strength, which can be many months depending on the conditions of moisture and temperature.

(3) Large shrinkage: A large amount of water evaporates during hardening, and the volume shrinks greatly due to the loss of capillary water. The porous structure also contributes to the large shrinkage, which may cause cracking.

(4) Poor water resistance: In a humid environment, the moisture in the lime can not evaporate, and the hardening stops. That's why lime is an air-hardening material. The hardened lime has poor resistance to water mainly because it is porous and the calcium hydroxide is soluble in water.

The quality of quicklime is mainly determined by the content of effective calcium oxide, magnesium oxide, over-burnt lime, and other impurities in lime. The carbon dioxide content can be tested based on the amount of calcium carbonate. The content of residues can be determined based on the residues on the 2.36 mm sieve in slaked lime. The specifications JC/T 479—2013 and JC/T 481—2013 classify quicklime and slaked lime used in buildings into three grades, as shown in Table 3.1 and Table 3.2 respectively. Table 3.3 shows the three grades of lime used in pavements according to specification JTG/T F20—2015.

Table 3.1 Grades of quicklime used in buildings (JC/T 479—2013)

Items	Calcitic quicklime			Dolomite quicklime	
	CL90	CL85	CL75	ML85	ML80
Min. CaO and MgO content (%)	90	85	75	85	80
MgO content (%)	≤5			>5	
Max. CO_2 content (%)	4	7	12	7	7
Max. SO_3 content (%)	2				
Min. volume of slurry (L/kg)	2.6			—	
Fineness — Max. retained on 0.2 mm sieve (%)	2			7	
Fineness — Max. retained on 0.09 mm sieve (%)	7			2	

Note: CL means CaO lime; ML means MgO lime.

Table 3.2 Grades of slaked lime used in buildings (JC/T 481—2013)

Items	Calcitic slaked lime			Dolomite slaked lime	
	HCL90	HCL85	HCL75	HML85	HML80
Min. CaO and MgO content (%)	90	85	75	85	80
MgO content (%)	≤5			>5	
Max. SO_3 content (%)	2				
Max. free water content (%)	2				
Volume stability	Qualified				
Fineness — Max. retained on 0.2 mm sieve (%)	2				
Fineness — Max. retained on 0.09 mm sieve (%)	7				

Note: HCL means hydrated CaO lime; HML means hydrated MgO lime.

Table 3.3 Grades of lime used in pavement engineering (JTG/T F20—2015)

Items	Calcitic quicklime			Dolomite quicklime			Calcitic slaked lime			Dolomite slaked lime		
	I	II	III	I	II	III	I	II	III	I	II	III
Min. effective CaO and MgO content (%)	85	80	70	80	75	65	65	60	55	60	55	50
Max. impurity (Residues on 5 mm sieve) content (%)	7	11	17	10	14	20	—	—	—	—	—	—

Continued

Items		Calcitic quicklime			Dolomite quicklime			Calcitic slaked lime			Dolomite slaked lime		
		I	II	III	I	II	III	I	II	III	I	II	III
Max. water content (%)		—	—	—	—	—	—	4	4	4	4	4	4
Fineness	Max. retained on 0.6 mm sieve (%)	—	—	—	—	—	—	0	1	1	0	1	1
	Max. retained on 0.15 mm sieve (%)	—	—	—	—	—	—	13	20	—	13	20	—
Content of CaO and MgO (%)		≤5			>5			≤4			>4		

3.2 Production of Cement

Cement is a fundamental material in civil engineering. 2000 years ago Romans started using lime and volcanic ash or pozzolan, which is a natural form of cement. In 1824, English inventor Joseph Aspdin invented Portland cement, which has remained the dominant form of cement used in concrete. In 1889, China built the first cement plant in Tangshan. Since 1985, China has been producing the majority of the cement in the world. In 2019, the world produced 4.4 billion tons of cement. China produced and consumed about 60% of the world's cement. The famous three Gorges Dam used 16 million tons of cement. The main application of Portland cement is to make Portland cement concrete. It can also be used for other purposes, such as stabilizing soil and aggregate base for highway construction. Understanding the production, chemical composition, hydration, and properties of Portland cement are critical for the selection, design, and quality control of cement concrete.

3.2.1 Classification

Based on the content of supplementary materials and the purposes, the specifications in China generally classify cement into four types:

(1) Portland cement is a hydraulic binding material mainly consisting of Portland cement clinker, 0-5% limestone or granulated blast furnace slag, and a limited amount of gypsum. According to specification GB 175—2007, Portland cement is further classi-

fied into two types: Type I Portland cement, coded as P·I, does not contain supplementary materials; Type II Portland cement, coded as P·II, contains ≤5% ground limestone or granulated blast furnace slag.

(2) The ordinary Portland cement contains 5%–20% supplementary materials.

(3) The specific or blended Portland cement which contains a higher percentage of supplementary materials includes fly ash cement, slag cement, and pozzolanic cement. The highest strength grade for ordinary Portland cement and blended Portland cement is 52.5, while that for Portland cement is 62.5.

(4) The special cement is the cement with special functions, such as oil well cement, pavement cement, quick-set cement, and expansive cement.

3.2.2 Production

As shown in Figure 3.3, the Portland cement production process begins with mixing and grinding the limestone and clay at specific proportions and then heating to a sintering temperature in a kiln to produce clinker which are rounded nodules between 1 mm and 25 mm. Then, the clinker is cooled and ground with gypsum into a fine powder in a mill to create cement.

The raw materials for producing Portland cement include lime, silica, alumina, and iron oxide, which can be classified into two types of ingredients: a calcareous material and an argillaceous material. The calcareous materials such as limestone, chalk, or oyster shells provide calcium oxide. The argillaceous materials such as clay, shale, and blast furnace slag provide silica, alumina, and ferrite oxides. These raw materials are crushed, stored in silos, and then ground in the desired proportions using either a wet or dry process. The ground material is stored and then heated and sent to the kiln, where these raw materials interact to form complex chemical compounds. Modern dry process cement plants use a heat recovery unit to preheat the ground materials or feedstock with the exhaust gas from the kiln. Some plants use a furnace to further heat the feedstock. Both the preheater and furnace improve the energy efficiency during cement production. The final process is to grind the clinker into a fine powder and a small amount of gypsum (calcium sulfate or $CaSO_4$) is added during grinding to regulate the setting time of the cement.

Different calcine temperatures lead to different mineral composition of cement. At 800 °C, dicalcium silicate (C_2S) starts to be generated. At 900 to 1100 °C, tricalcium aluminate (C_3A) and tetracalcium aluminoferrite (C_4AF) are generated. At 1200 to

1300 ℃ the liquid phase is formed. At 1100 to 1200 ℃, a large number of C_2S, C_3A, and C_4AF are generated. At 1300 to 1450 ℃, C_2S synthesizes C_3S by absorbing calcium oxide (CaO), which is the key to the production of cement.

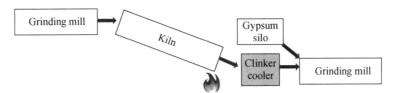

Figure 3.3 The production of Portland cement

3.2.3 Composition

Cement has four main compounds including C_3S, C_2S, C_3A, and C_4AF. Table 3.4 summarizes the formula, abbreviate, content, reaction rate, reaction heat, and strength contribution of the four main compounds. Generally, C_3S accounts for 36%–60% of cement and has a moderate reaction rate and reaction heat. C_2S accounts for 15%–37% of cement and has a slow reaction rate and low reaction heat. C_3A accounts for 7%–15% of cement and has a fast reaction rate and very high reaction heat. C_4AF accounts for 10%–18% of cement and has a moderate reaction rate and reaction heat. C_3S is the most abundant compound in cement clinker, and it hydrates more rapidly than C_2S contributing to the final setting time and early strength gain of the cement paste.

Table 3.4 Properties of main compounds of Portland cement

Compound	Formula	Abbreviation	Content	Reaction rate	Reaction heat	Strength contribution
Tricalcium Silicate	$3CaO \cdot SiO_2$	C_3S	36%–60%	Moderate	Moderate	Early and long-term
Dicalcium Silicate	$2CaO \cdot SiO_2$	C_2S	15%–37%	Slow	Low	Long-term
Tricalcium Aluminate	$3CaO \cdot Al_2O_3$	C_3A	7%–15%	Fast	Very high	Low
Tetracalcium Aluminoferrite	$4CaO \cdot Al_2O_3 \cdot Fe_2O_3$	C_4AF	10%–18%	Moderate	Moderate	Bending

The properties of cement vary based on the proportions of the main compounds. Figure 3.4 shows the strength of the four main compounds over the curing time. C_3S provides both early and long-term strength because of its moderate reaction rate and large amount. C_2S only provides long-term strength due to the slow reaction rate. C_3A and C_4AF provide low strength due to the limited amount, although the reaction rate of C_3A is fast. C_4AF mainly contributes to the bending strength of cement because its hydration product is a needle-shaped crystal that acts as reinforcement in the hardened cement.

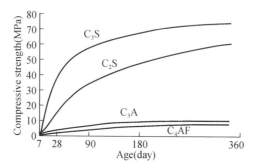

Figure 3.4 Strength growth curve of cement clinker during hardening

In addition to these four main compounds, Portland cement includes several minor compounds, including free calcium oxide (f-CaO), free magnesium oxide (f-MgO), and alkaline minerals such as sodium oxide (Na_2O) and potassium oxide (K_2O), etc. These minor compounds represent a few percentages by weight of cement. The term minor compounds refer to their quantity and not to their importance. The alkaline minerals can react with the active silica in some aggregates causing the disintegration of concrete and their content should be controlled at a low or safe level. Figure 3.5 illustrates the compounds in a Portland cement particle. C_3S is the light color angular crystal. C_2S is the dark round crystal. C_3A is the needle-shaped crystal. C_4AF is the smallest crystal. The C_3A and C_4AF occupy interstitial space between C_3S and C_2S. In addition, we have a small amount of f-CaO and f-MgO crystals.

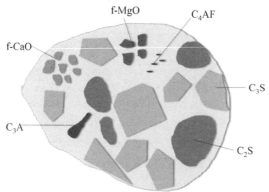

Figure 3.5 Compounds in a Portland cement particle

3.3 Hydration of Cement

3.3.1 Hydration Products

Hydration is the chemical reaction between cement particles and water. This reaction includes the change of matter and molecule energy level. When the cement particles contact water, the minerals in the cement clinker immediately hydrate with water and generate new hydration products and release heat. The main hydration products include calcium silicate hydrate ($3CaO \cdot 2SiO_2 \cdot 3H_2O$ or CSH), calcium hydroxide ($Ca(OH)_2$ or CH), calcium aluminate hydrate ($3CaO \cdot Al_2O_3 \cdot 6H_2O$ or CAH), calcium ferrite hydrate ($CaO \cdot Fe_2O_3 \cdot H_2O$ or CFH), and calcium sulfoaluminate hydrate ($3CaO \cdot Al_2O_3 \cdot 3CaSO_4 \cdot 31H_2O$ or CASH). Below are the chemical reactions during hydration and the abbreviated form. In summary, C_3S and C_2S react with water to form CSH and CH. C_3A reacts with water to form CAH. C_4AF reacts with water to form CAH and CFH. CAH then reacts with water and calcium sulfate (CS) to form CASH.

$$2(3CaO \cdot SiO_2) + 6H_2O = 3CaO \cdot 2SiO_2 \cdot 3H_2O + 3Ca(OH)_2$$
$$2(2CaO \cdot SiO_2) + 4H_2O = 3CaO \cdot 2SiO_2 \cdot 3H_2O + Ca(OH)_2$$
$$3CaO \cdot Al_2O_3 + 6H_2O = 3CaO \cdot Al_2O_3 \cdot 6H_2O$$
$$4CaO \cdot Al_2O_3 \cdot Fe_2O_3 + 7H_2O = 3CaO \cdot Al_2O_3 \cdot 6H_2O + CaO \cdot Fe_2O_3 \cdot H_2O$$
$$3CaO \cdot Al_2O_3 \cdot 6H_2O + 3(CaSO_4 \cdot 2H_2O) + 19H_2O =$$
$$3CaO \cdot Al_2O_3 \cdot 3CaSO_4 \cdot 31H_2O$$
$$C_3S + H \longrightarrow CSH \text{ (gel)} + CH \text{ (hexagon crystal)}$$
$$C_2S + H \longrightarrow CSH \text{ (gel)} + CH \text{ (hexagon crystal)}$$
$$C_3A + H \longrightarrow CAH \text{ (cubic crystal)}$$
$$C_4AF + H \longrightarrow CAH \text{ (cubic crystal)} + CFH \text{ (gel)}$$
$$CAH + CS + H \longrightarrow CASH \text{ (needle crystal)}$$

Table 3.5 summarizes the properties of the main hydration products. Figure 3.6 shows the microscopic image of the main hydration products. The main hydration product is CSH. It is an insoluble gel-like poorly crystalline material, accounts for 70% of the hydration products, and is primarily responsible for the strength of concrete. CSH is not a well-defined compound since the calcium-to-silicate ratio varies between 1.5 and 2, and the structurally combined water content is more variable. The rate of hydration is

Table 3.5 The properties of the main hydration products

Product	Abbreviation	Form	Solubility	Proportion
Calcium silicate hydrate	CSH	Gel	Insoluble	70%
Calcium hydroxide	CH	Hexagon	Soluble	20%
Calcium aluminate hydrate	CAH	Cubic	Soluble	—
Calcium ferrite hydrate	CFH	Gel	Insoluble	—
Calcium sulfoaluminate hydrate (Ettringite)	CASH	Needle	Slightly soluble	7%

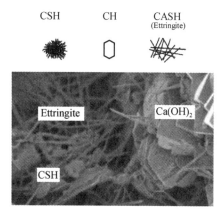

Figure 3.6 The microscopic image of the main hydration products

accelerated by sulfate ions in the solution. Thus, a secondary effect of the addition of gypsum to cement is to increase the rate of development of the CSH. Complete hydration of C_3S produces 61% CSH and 39% CH; hydration of C_2S results in 82% CSH and 18% CH.

CH is a hexagonal crystal accounting for 20% of the hydration products. It is easily soluble in water and therefore the lime concentration of the solution quickly reaches the saturation state. The hydration of cement is mainly carried out in the saturated solution of lime.

The reaction of C_3A is the fastest and releases large amounts of heat. CS can react with C_3A and slow down the rate of aluminate hydration. CAH is a soluble cubic crystal and CFH is an insoluble gel-like material. CASH, also called ettringite, is a needle-shaped crystal slightly soluble in water, and accounts for 7% of the hydration products.

The aluminate hydrates much faster than the silicate. The reaction of C_3A with water is immediate and releases large amounts of heat. CS is used to slow down the rate of aluminate hydration. The CS dissolves into the solution quickly, producing sulfate ions that suppress the solubility of the aluminate. The balance of aluminate to sulfate is critical for the rate of setting.

(1) If the availability of both aluminate and sulfate ions is low, the cement paste will remain flowable for about 45 minutes. Then, the paste starts to harden as crystals displace the water in the pores and begin to solidify within 2 to 4 hours.

(2) If there is an excess of both aluminate and sulfations ions, the workable stage may only last for 10 minutes and hardening may occur in 1 to 2 hours.

(3) If the availability of aluminate ions is high, and sulfates are low, either a quick set (10 to 45 minutes) or flash set (less than 10 minutes) can occur.

(4) If the availability of aluminate ions is low and the availability of sulfate ions is high, the CS can recrystallize in the pores within 10 minutes, producing a flash set.

3.3.2 Hydration Process

Hydration is the reaction of the cement with water to produce a hydrate, which brings about chemical, physical and mechanical changes in the system. There are two mechanisms for the hydration of Portland cement, the through-solution reaction and topochemical hydration. The through-solution reaction dominates the early stages of hydration and involves the dissolution of anhydrous compounds into the solution, the formation of hydrate products in the solution, and the precipitation of hydrate products from the supersaturated solution. The topochemical hydration is a solid-state chemical reaction occurring at the surface of the cement particles. Some studies found that low water/solid ratios lead to a topochemical reaction, while the through-solution reaction is more important at high water/solid ratios, and both reactions can occur simultaneously.

After mixing cement with water, the cement paste has plasticity but will gradually lose plasticity, which is called setting. Then, the cement paste becomes hard and gains strength, which is called hardening. Cement setting and hardening is a continuous and complex process of physical and chemical changes, which determines a series of technical properties of cement. Scholars have researched this process for more than one hundred years and still hold different opinions. Usually, the initial setting is the time when cement can be molded in any desired shape without losing its strength. The final setting is the time when cement completely loses its plasticity and becomes hard. Hardening means cement paste gains strength through hydration of silicate compounds. The mechanism of setting and hardening can be explained with the sequential development of the structure in a cement paste during hydration as summarized in Figure 3.7, which generally includes four stages:

(1) During the early stages of hydration, weak bonds can form, particularly from the gel of hydrated C_3A.

(2) The initial set occurs when further hydration stiffens the mix and begins locking the structure of the material in place.

(3) The final set occurs when the CSH has developed a rigid structure, and all components of the paste are locked into place.

(4) The space between the cement grains is filled with hydration products. The cement paste continues hardening and gains strength as hydration continues.

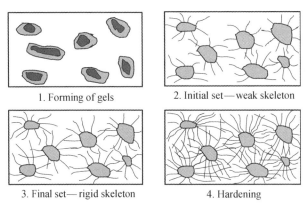

1. Forming of gels 2. Initial set—weak skeleton
3. Final set—rigid skeleton 4. Hardening

Figure 3.7 Hydration process

3.3.3 Influencing Factors

The cement setting and hardening process are not only affected by the compounds of the cement, but also by many other factors.

1) Fineness

As shown in Figure 3.8, finer cement particles increase surface specific area and therefore increase the hydration activity of cement. When the particle size is larger than 45 μm, the hydration is difficult, and when the particle size is larger than 75 μm, there is almost no hydration.

3 Inorganic Binding Materials

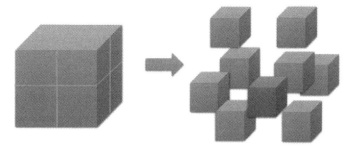

Figure 3.8 Increased specific surface area for finer cement particles

2) Gypsum

As shown in Figure 3.9, gypsum ($CaSO_4$ or CS) can react with CAH quickly to generate CASH which deposits and forms a protection film on the cement particles to hinder the hydration of C_3A and delay the setting time of cement. When there is sufficient CS, we have calcium tri-sulfoaluminate hydrate, also called Aft and its natural form is ettringite. When CS content is low, we have calcium mono-sulfoaluminate hydrate, also called Afm. Here, A means alumina, f means ferric oxide, t means tri, and m means mono.

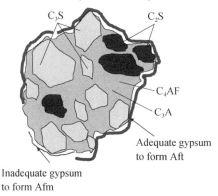

Figure 3.9 Effects of gypsum in cement hydration

3) Curing time

The amount of hydration products increases with the increase of time and therefore strength increases. The curing time can be classified into setting and hardening. As shown in Figure 3.10, the amount of CSH increases rapidly in the first 28 days and then flats out. Therefore, the 28-day strength is usually tested for determining cement and concrete strength.

Civil Engineering Materials

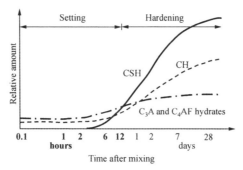

Figure 3.10 Relative amount of hydration products over curing time

4) Water-cement ratio

Figure 3.11 shows that with the increase of water-cement ratio, the proportional volume of cement hydrates increases first and then decreases. When the water-cement ratio is lower than 0.38, there is not sufficient water to hydrate all cement particles and unhydrated cement remains. However, if the water-cement ratio is too high, there will be a large number of capillaries in the hardened cement paste, increasing porosity and reducing strength.

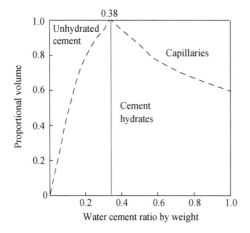

Figure 3.11 The relationship between proportional volume and water-cement ratio

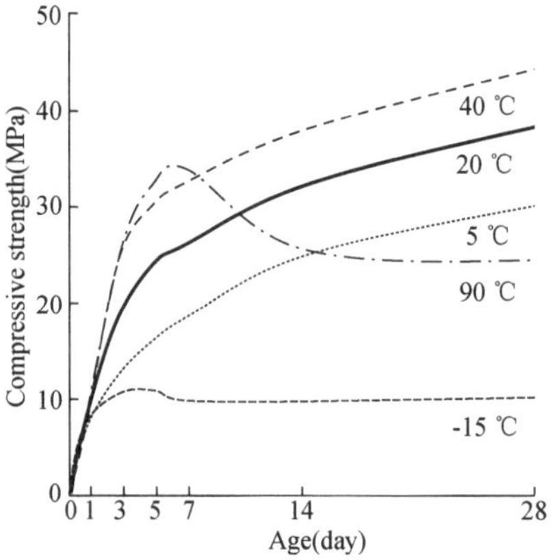

Figure 3.12 Compressive strength different curing temperatures

5) Temperature

As shown in Figure 3.12, high temperature accelerates hydration and improves the 28-day strength. However, if the temperature is too high, for example, close to 100 degrees centigrade, the hydration is disturbed and the strength is very low. Low temperature causes the reduction of strength. When water is frozen, hydration nearly stops.

6) Moisture

Hydration needs sufficient water. The curves in Figure 3.13 show the compressive strength of cement under different moist curing times. The longer the moisture curing time, the higher the strength. A dry environment causes fast evaporation. When water evaporates, the hydration stops causing shrinkage cracks and low strength.

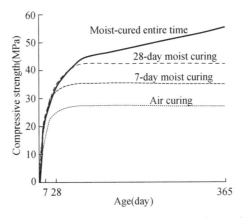

Figure 3.13 Compressive strength at different curing moisture levels

3.4 Properties of Cement

Cement is mainly used for the preparation of mortar and concrete in civil engineering. The proper hydration of Portland cement is a fundamental quality for cement producers, whereas the quality of the concrete is also influenced by mix proportion, characteristics of aggregates, and quality control. Properties of the hydrated cement are evaluated with either cement paste which is the mix of water and cement, or mortar which is the mix of water, cement, and sand. The specification GB 175—2007 provides the requirements of the chemical and physical properties of cement. In addition to the properties such as fineness, setting time, soundness, and strength, it also has requirements of the amount of insoluble impurities, the loss during sintering, the content of SO_3, MgO, and chloride ion, and fineness.

3.4.1 Density

The density and unit weight of cement are often needed in the calculation of the proportions in concrete, storing, and transporting. The density of Portland cement is generally 3–3.15 g/cm³. The unit weight mainly depends on the packing of cement, in addition to the mineral composition and fineness. The unit weight of slack cement is around 1–1.1 g/cm³, while that of compact cement is around 1.6 g/cm³.

3.4.2 Fineness

Fineness refers to the particle size of cement. The fineness of cement particles is an important property that must be carefully controlled. The finer the cement particle, the larger the surface area and the faster the hydration. Therefore, finer material results in faster strength development and greater initial heat of hydration. Cement particles larger than 0.045 mm will have hydration difficulties. 85% to 95% of the cement particles are smaller than 0.045 mm, and the average diameter is 0.01 mm. Figure 3.14 shows the irregular shapes of cement particles at the micro-scale.

The fineness of cement can be evaluated by the specific surface area, which is the total surface area per unit weight of cement particles, and can be measured with the Blaine air permeability apparatus. In the specification GB 175—2007, the specific surface area of Portland cement and ordinary Portland cement must be greater than 300 m²/kg. For the blended cement including slag cement, fly ash cement, Pozzolanic cement, etc., the percentage retained at 0.08 and 0.045 mm sieve should be no more than 10% and 30%, respectively.

(a) 100 μm scale (b) 10 μm scale

Figure 3.14 Scanning electron microscope (SEM) image of cement particles

3.4.3 Consistency

Water content for standard consistency refers to the amount of water, expressed as a percentage of the weight of cement, required to mix cement into a specific plastic state (which is called the standard consistency or normal consistency). The properties of the cement paste are tested at the standard consistency so that the results can be compared. In the specification GB/T 1346—2011, the standard consistency allows a standard plunger of 10 mm diameter to penetrate to 6 mm ± 1 mm above the bottom of the Vicat mold (Figure 3.15). During the test, the 1 mm diameter needle is allowed to penetrate the paste for 30 seconds and the amount of penetration is measured. The water content at the standard consistency of Portland cement is generally between 24% and 30%, mainly determined by the mineral composition and fineness.

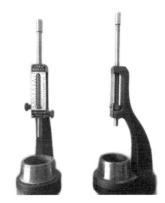

Figure 3.15　Consistency tester

3.4.4 Setting Time

When we add water to cement, the cement paste changes from a fluid to a plastic solid and then a hardened solid. Setting refers to the stiffening of the cement paste or the change from a plastic state to a solid state and is classified into the initial setting and final setting. The initial setting is the time that elapses from the moment water is added until the paste ceases to be fluid and plastic. The final setting is the time required for the paste to acquire a certain degree of hardness, lose all plasticity/flowability and start to gain strength. The setting time of cement is important for construction. The initial setting time should not be too fast to have enough time for mixing, transportation, and casting; whereas, after the completion of the casting, the concrete should harden quickly.

Figure 3.16 shows the definition of setting time. The setting time of cement is tested with the cement paste of standard consistency under specified temperature and humidity using the consistency tester. In the specification GB/T 1346—2011, the initial

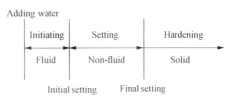

Figure 3.16　Definition of setting time

setting time should be tested from the time when cement is added into water to the time when the needle sinks to 4 mm ±1 mm from the bottom plate in the neat paste. The final setting time should be tested from the time when cement is added into the water to the time when the needle sinks into the neat paste 0.5 mm without a ring mark. According to specification GB 175—2007, for Portland cement, the initial setting time should be no less than 45 minutes, and the final setting time should be no more than 390 minutes. For other types of cement, the initial setting time should be no less than 45 minutes.

3.4.5 Soundness

The soundness of the cement paste refers to its ability to retain its volume after setting. Expansion after setting is mainly caused by delayed or slow hydration or other reactions, causing expansive cracks and poor quality of the structure. According to specification GB/T 1346—2011, the soundness can be evaluated by the pat test and the Le Chatelier test. The pat test is to prepare a cement cake with cement paste of standard consistency, boil it for 3 h, and then observe it with naked eyes. If there is no crack and no bending by ruler inspection, it is called qualified soundness. The Le Chatelier test is to measure the expansion value after the cement paste is boiled and get hardened on Le Chatelier needles. If the expansion value is within the required value (<5 mm), its stability is qualified. If there is a contradiction between the results of the two tests, the Le Chaterlier test shall prevail.

Too much f-CaO and f-MgO are the main reasons for poor volume stability. Excessive CS can also easily lead to volume expansion, because the sulfur trioxide ions in CS will react with CAH to form CASH which leads to excess volume expansion. The poor stability caused by f-CaO is tested by boiling curing. The poor stability caused by f-MgO must be tested by the pressure steam curing, because the hydration of f-MgO is slower than that of f-CaO. Generally, the content of f-CaO, f-MgO, and SO_3 ions must be strictly restricted.

3.4.6 Strength

Strength is the key technical performance for selecting cement. The strength mainly depends on the mineral composition and fineness of the cement. Since the strength increases gradually in the hardening process, the strength is evaluated at different curing ages. The specification GB/T 17671—2021 introduces a method for testing the three-point flexural strength of the cement mortar beam and the compressive strength of each fractured part. The cement, standard sand, and water are mixed at a ratio of 1:3:0.5 to prepare the 40 mm × 40 mm × 160 mm cement mortar beams which will be cured under the standard condition of 20 ℃ ±1 ℃ and 90% humidity for 3 and 28 days.

The loading rate for the flexural test is 50 N/s and the flexural strength is calculated by Equation (3.4). The final flexural strength is the average of three specimens. If the result of any specimen is not within ±10% of the average of the three, that result should be dropped, and the final result is the average of the rest two specimens. If any of the two values is not within ±10% of the average value, the test result is invalid.

$$R_f = \frac{3F_f L}{2bh^2} = 0.00234 F_f \tag{3.4}$$

where R_f = flexural strength (MPa);

F_f = failure load (N);

L = distance between two supporting points (100 mm);

b = width of the beam (40 mm);

h = height of the beam (40 mm).

The fractured specimens from the flexural tests should be tested immediately for compressive strength. The top and bottom compression surfaces are the two lateral sides of the cement beam. The loading rate is 2400 N/s and the compressive strength is calculated by Equation (3.5). The final compressive strength is the average of the six specimens. If any of the specimens is not within ±10% of the average of the six, that result shall be dropped, and the final result is the average of the rest five specimens. If any of the five values is not within ±10% of the average value, the test result is invalid.

$$R_c = \frac{F_c}{A} \tag{3.5}$$

where R_c = compressive strength (MPa);

F_c = failure load (N);

A = area of compression surface, 40 mm × 40 mm = 1600 mm^2.

According to specification GB 175—2007, the Portland cement can be classified into three strength grades including 42.5, 52.5, and 62.5, according to the 3-day and 28-day flexural strength and compressive strength. If the cement has higher flexural strength and compressive strength at an early age, it is called early-strength cement, which comes with a serial number "R". The strength at different ages of cement of each strength grade should not be lower than the values in Table 3.6.

Table 3.6 Strength grades of different cement

Types	Grades	Compressive strength (MPa)		Flexural strength (MPa)	
		3-day	28-day	3-day	28-day
Portland cement	42.5/42.5R	≥17/≥22	≥42.5	≥3.5/≥4	≥6.5
	52.5/52.5R	≥23/≥27	≥52.5	≥4/≥5	≥7
	62.5/62.5R	≥28/≥32	≥62.5	≥5/≥5.5	≥8
Ordinary Portland cement	42.5/42.5R	≥17/≥22	≥42.5	≥3.5/≥4	≥6.5
	52.5/52.5R	≥23/≥27	≥52.5	≥4/≥5	≥7
Slag, pozzolanic, fly ash, and blended cement	32.5/32.5R	≥10/≥15	≥32.5	≥2.5/≥3.5	≥5.5
	42.5/42.5R	≥15/≥19	≥42.5	≥3.5/≥4	≥6.5
	52.5/52.5R	≥21/≥23	≥52.5	≥4/≥5	≥7

3.4.7 Hydration Heat

Hydration heat refers to the heat released in the hydration process of cement. Most of the hydration heat is released at the initial stage within 7 days and then decreases gradually. Heat develops rapidly during setting and initial hardening and gradually declines and finally stabilizes as hydration slows. Figure 3.17 shows the hydration heat over curing time, which includes three peaks. The first peak is during the initial wetting, mainly caused by the formation of ettringite (Aft) of C_3A. The second peak is during the hydration of C_3S, the third peak is during the transformation from Aft to Afm, and the hydration of C_3A. The hydration heat is determined by the mineral composition and fineness of the cement. In winter construction, the hydration heat is beneficial for the cement hardening. But for mass concrete structures, too much hydration heat is harmful. The hydration heat accumulated in the interior is not easy to release and the internal temperature can reach 70 − 90 ℃, resulting in temperature stress and cracks.

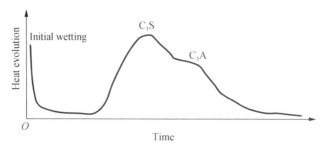

Figure 3.17 Cement hydration heat over curing time

3.4.8 Voids

Due to the random growth of the crystals and the different types of crystals, voids are left in the paste structure as cement hydrates and have a great influence on the strength, durability, and volume stability of concrete. Two types of voids are formed during hydration: the interlayer hydration space and capillary voids. In addition to the interlayer hydration space and capillary voids, the cement paste contains trapped air which reduces strength and increases permeability. However, well-distributed small air bubbles can greatly increase the durability of the cement paste.

Interlayer hydration space is between the layers in the CSH. The space thickness is $0.5-2.5$ nm, which is too small to influence the strength but contributes 28 % to the porosity. Water in the interlayer hydration space is held by hydrogen bonds, but can be removed when humidity is less than 11 %, causing shrinkage. Capillary voids are mainly determined by the spaces and hydration of cement particles. For a highly hydrated cement paste with a small amount of water, the capillary voids are $10-50$ nm. A poorly hydrated cement produced with excess water has capillary voids of 3-5 μm. Capillary voids greater than 50 nm decrease strength and increase permeability. Removal of water from capillary voids greater than 50 nm does not cause shrinkage, whereas removal of water from the smaller voids causes shrinkage.

3.5 Corrosion

Portland cement has good durability in a normal environment after hardening. Studies show that the compressive strength of cement concrete after 30 to 50 years can be 30 % higher than that of 28 days. However, corrosion may occur when the cement hydration products react with other chemicals in the environment resulting in surface chalking, discoloration, cracking, steel rusting, and other distress. Generally, corrosion is mainly caused by either the formation of expansive products or the loss of materials due to the dissolution of hydration products in water or the reaction of hydration products with acids or ions.

3.5.1 Soft Water Corrosion

Soft water refers to water with small temporary hardness, such as rainwater, snow water, river water, and lake water containing less bicarbonate. Hard water is formed when water percolates through deposits of limestone, chalk, or gypsum which are largely made up of calcium and magnesium carbonate, bicarbonate, and sulfate. Temporary hardness is calculated as the bicarbonate content per liter of water.

When the cement mortar contacts soft water for a long time, some of the hydration products will be gradually dissolved in water. CH has the greatest solubility about 1.2 g/L at 25 ℃. In the case of still water or with no water pressure, the surrounding water is rapidly saturated with the dissolved CH. The dissolution effect is quickly terminated. Therefore the dissolution is limited to the surface and has little effect. But in flowing water, especially under water pressure, water will continuously dissolve CH and carry it away. Other hydration products, such as CAH, CAH, etc., may also dissolve in water, which will reduce the strength, and even cause the failure of the structure. The surface corrosion is shown in Figure 3.18(a). The higher the temporary hardness of water, the weaker the corrosion of cement mortar. Because the bicarbonate in water can react with CH in cement to form $CaCO_3$ which is insoluble in water, as shown in Equation (3.6). The $CaCO_3$ accumulates in the pores of the cement mortar, forming a dense protective layer, and preventing the infiltration of water.

$$Ca(OH)_2 + Ca(HCO_3)_2 \Longleftrightarrow 2CaCO_3 + 2H_2O \tag{3.6}$$

3.5.2 Sulfate Attack

The seawater, salt marsh water, groundwater, and some industrial sewage may contain sulfate, which has a corrosive effect on cement mortar. Sulfate reacts with calcium hydrates to form CS and then reacts with CAH to form CASH, as shown in Equations (3.7) and (3.8). CASH contains a large amount of crystal water, and its volume increases by 1.5 times. CASH is often called "cement bacillus" because it is a needle-shaped crystal. This reaction has a destructive effect on cement concrete, as shown in Figure 3.18(b). Adding the CS to the cement clinker to regulate the setting time can also produce CASH, but its content is strictly restricted and will not cause damage.

$$2H^+ + SO_4^{2-} + Ca(OH)_2 \Longleftrightarrow CaSO_4 \cdot 2H_2O \tag{3.7}$$

$$3CaO \cdot Al_2O_3 \cdot 6H_2O + (3CaSO_4 \cdot 2H_2O) + 19H_2O \Longleftrightarrow 3CaO \cdot Al_2O_3 \cdot 3CaSO_4 \cdot 27H_2O \tag{3.8}$$

3.5.3 Magnesium Corrosion

Seawater and groundwater often contain a large number of magnesium salts, mainly magnesium sulfate and magnesium chloride. They usually have a replacement reaction with CH. Magnesium salts react with calcium hydroxide to form calcium dichloride ($CaCl_2$) and calcium sulfate ($CaSO_4$, i.e. CS) as shown in Equations (3.9) and (3.10). Because magnesium hydroxide is soft and unbonded, and calcium chloride is soluble in water, magnesium corrosion can seriously damage the structure, especially if there is also sulfate in the water. CS can react with CAH to form CASH. The magnesium corrosion is shown in Figure 3.18(c).

$$MgCl_2 + Ca(OH)_2 \Longleftrightarrow CaCl_2 + Mg(OH)_2 \quad (3.9)$$

$$2H_2O + MgSO_4 + Ca(OH)_2 \Longleftrightarrow CaSO_4 \cdot 2H_2O + Mg(OH)_2 \quad (3.10)$$

(a) Soft water corrosion

(b) Sulfate attack

(c) Magnesium corrosion

Figure 3.18 Typical corrosion of cement concrete

3.5.4 Carbonation

In most natural water, there is usually some free carbon dioxide and its salts. If the free carbon dioxide is too much, it will be destructive. Calcium hydroxide reacts with carbon dioxide and water to form calcium carbonate which is low bonding, or calcium hydrogen carbonate which is soluble, as shown in Equations (3.11) and (3.12). Calcium hydroxide in cement mortar is dissolved by transforming into soluble calcium hydrogen carbonate, leading to the decomposition of other hydrates, and enhancing the corrosion effect.

$$Ca(OH)_2 + CO_2 + H_2O = CaCO_3 \cdot 2H_2O \qquad (3.11)$$
$$CaCO_3 + CO_2 + H_2O = Ca(HCO_3)_2 \qquad (3.12)$$

3.5.5 Acid Corrosion

Various acids are often found in industrial waste water, groundwater, and marsh water which have a corrosion effect on cement mortar. The acids react with calcium hydroxide to form compounds that either dissolve in water or expand in volume, resulting in destruction. Typical acid corrosion includes hydrochloric and sulfuric acid corrosion. Hydrochloric acid reacts with calcium hydroxide to form calcium chloride which is soluble in water, as shown in Equation (3.13). Sulfuric acid (H_2SO_4) reacts with CH to form CS, as shown in Equation (3.14). The CS either leads to volume expansion or reacts with CAH to form CASH causing cracking. Generally, the greater the concentration of hydrogen ions in water, the smaller the pH, the more severe the corrosion.

$$2HCl + Ca(OH)_2 = CaCl_2 + 2H_2O \qquad (3.13)$$
$$H_2SO_4 + Ca(OH)_2 = CaSO_4 \cdot 2H_2O \qquad (3.14)$$

The above types of corrosion can be summarized into the following three types of damage:

(1) Dissolution corrosion: Some media gradually dissolve components of cement mortar, resulting in dissolution damage.

(2) Ion exchange: Ion exchange reaction occurs between corrosion materials and cement components, and forms dissolution products or unbonded products which can destroy the original structure.

(3) Forming expansive compounds: The salt crystals formed by corrosion reactions increase in volume and generate harmful internal stress, causing cracking or other expansion failures.

3.5.6 Measures

According to the different corrosion causes, the following prevention measures can be taken to reduce the risk of corrosion:

(1) Choosing proper cement according to the environment: If the cement mortar suffers from soft water, it is recommended to select the cement with less calcium hydroxide content. If the cement mortar suffers from sulfate corrosion, it is recommended to select sulfate-resistant cement with low content of tricalcium aluminate. Additionally, there are some types of Portland cement clinker added with artificial or natural mineral materials (mixed materials), which have better corrosion resistance.

(2) Reducing permeability: Improving the impermeability of cement mortar is an effective measure to prevent corrosion. The greater the density of the cement mortar, the stronger the impermeabilty, the more difficult for the corrosive media to enter. Reducing permeability also can resist soft water corrosion.

(3) Using protection sealing: When taking the above measures is still difficult to prevent corrosion, it is recommended to apply a waterproof layer with strong corrosion resistance on the surface of cement products. Common coatings include acid-resistant stone, acid-resistant ceramics, glass, plastics, asphalt, etc.

To control cement quality, there are some properties of cement should be tested. Insoluble residues are to evaluate the insoluble residues in HCl or $NaCO_3$ solutions to check the impurities in cement. Loss on ignition is to check the weight loss of cement heated at 950 ℃ for 15 minutes. As the main cause of steel corrosion, the chloride ion content must be tested. According to specification GB 175—2007, the chloride ion in cement should be less than 0.06%. To control the content of f-MgO and SO_3 in cement, the soundness should be also tested.

3.6 Supplementary Materials and Blended Cement

3.6.1 Supplementary Materials

Since the 1970s, some byproducts of other industries have been used in cement as supplementary cementitious materials. These supplementary cementitious materials mainly include fly ash, slag, and pozzolan. Since these materials are cementitious, they can be used in addition to or as a partial replacement for Portland cement. They are added as part of the total cementitious system to adjust the strength, improve cement performance and reduce the cost. The materials usually contain active SiO_2 and Al_2O_3, which can react with CH to form CSH or CAH, and CAH can further react with gypsum to form

CASH, as shown in Equations (3.15)–(3.17). This is called the pozzolanic property or secondary hydration because it utilizes and relies on the CH which is generated during cement hydration. The key to the pozzolanic property is that the structure of the silica must be in a glassy or amorphous form with a disordered structure, which can be formed by rapid cooling from a molten state. Inter-molecular bonds in the structure are not at their preferred low energy orientation and are easy to break and link with calcium hydroxide.

$$x\text{Ca}(\text{OH})_2 + \text{SiO}_2 + n\text{H}_2\text{O} = x\text{CaO} \cdot \text{SiO}_2 \cdot (x+n)\text{H}_2\text{O} \quad (3.15)$$

$$y\text{Ca}(\text{OH})_2 + \text{Al}_2\text{O}_3 + m\text{H}_2\text{O} = y\text{CaO} \cdot \text{Al}_2\text{O}_3 \cdot (y+m)\text{H}_2\text{O} \quad (3.16)$$

$$\text{Al}_2\text{O}_3 + 3\text{Ca}(\text{OH})_2 + 3(\text{CaSO}_4 \cdot 2\text{H}_2\text{O}) + 23\text{H}_2\text{O} = 3\text{CaO} \cdot \text{Al}_2\text{O}_3 \cdot 3\text{CaSO}_4 \cdot 32\text{H}_2\text{O} \quad (3.17)$$

1) Slag

Slag, also called ground granulated blast-furnace slag (GGBFS), is a glass material formed by molten slag produced in the blast furnace as an industrial byproduct of the production of iron. The slag is developed in a molten condition simultaneously with iron in a blast furnace. The molten slag is rapidly chilled by quenching in water to form a glassy, sand-like granulated material. The material is then ground to less than 45 μm to be the GGBFS which contains CaO, MgO, Al_2O_3, SiO_2, Fe_2O_3, and other oxides and a small amount of sulfate. Generally, the CaO, SiO_2, and Al_2O_3 account for more than 90% of slag. Its chemical composition is similar to that of Portland cement. The quality of GGBFS is evaluated by the quality coefficient (K) as shown in Equation (3.18). The quality coefficient is the weight ratio of active compounds over the less and inactive compounds in the slag. The higher the quality coefficient, the higher the activity of slag. The specification GB 175—2007 requires that the quality coefficient of GGBFS used for cement shall not be less than 1.2.

$$K = \frac{\text{CaO} + \text{MgO} + \text{Al}_2\text{O}_3}{\text{SiO}_2 + \text{MnO} + \text{TiO}_2} \quad (3.18)$$

2) Fly ash

Fly ash is the most commonly used supplementary material in cement. It is a byproduct of burning pulverized coal in electric power generating plants. China relies heavily on coal-fired power and generates around 600 million tons of fly ash in 2020. Combusting pulverized coal in an electric power plant burns off the carbon and most volatile materials. However, the carbon content of common coal ranges from 70% to 100%. The impurities such as clay, feldspar, quartz, and shale fuse as they pass through the combustion chamber. When those impurities fuse in suspension and are carried away from the combustion chamber by the exhaust gas, they cool and solidify into 1−50 μm solid and hollow spherical glassy fly ash. The particle diameters of fly ash range from 1 μm to more than 0.1 mm, with an average of 0.015 to 0.02 mm, and are 70% to 90% smaller than 45 μm. Fly ash is primarily silica glass composed of CaO, SiO_2, Al_2O_3, and Fe_2O_3. Fly ash can be classified into two types: Class F and Class C, depending on the content of CaO. Class F fly ash usually has less than 5% CaO but may contain up to 10%. Class C fly ash has 15% to 30% CaO. According to specification GB/T 1596—2017, the fly ash used in cement should meet the requirements in Table 3.7.

Table 3.7 Technical requirements for fly ash used in cement

Items	Types of fly ash	Requirements
Max. loss on ignition (%)	Class F and Class C	8.0
Max. water content (%)	Class F and Class C	1.0
Max. SO_3 content (%)	Class F and Class C	3.5
Max. f-CaO content (%)	Class F	1.0
	Class C	4.0
Max. soundness (mm)	Class C	5.0
Min. strength activity index (%)	Class F and Class C	70.0

3) Pozzolan

Pozzolan is a siliceous and aluminous material with little or no cementitious effect but in the presence of moisture, it can react with CH in a finely divided form to form compounds possessing cementitious properties. Natural pozzolan includes volcanic ash, tuff, pumice, zeolite, diatomite, etc. Artificial pozzolan includes calcining shale and clay, cinder, siliceous slag, etc.

3.6.2 Blended Cement

Supplementary materials can be added to Portland cement to replace part of cement to produce blended cement. The properties of Portland cement, Portland ordinary cement, Portland slag cement, Portland fly ash cement, and Portland pozzolanic cement, are shown in Table 3.8.

Table 3.8　Types of blended cement

Abbreviations	Names	Blended materials	Content (%)
P·I, P·II	Portland cement	Supplementary materials	<5
P·O	Portland ordinary cement	Supplementary materials	6-15
P·S	Portland slag cement	Slag	20-70
P·F	Portland fly ash cement	Fly ash	20-40
P·P	Portland Pozzolanic cement	Pozzolan	20-50

Table 3.9　Properties of blended cement

Properties	Portland cement	Portland ordinary cement	Portland slag cement	Portland fly ash cement	Portland Pozzolanic cement
Content (%)	<5	6-15	20-70	20-40	20-50
Workability	Moderate	Moderate	Moderate	Good	Moderate
Hydration	Very fast	Fast	Slow	Slow	Slow
Hydration heat	High	High	Low	Low	Low
Early strength	Very high	High	Low	Low	Low
Impermeability	Moderate	Moderate	Good	Good	Good
Corrosion resistance	Moderate	Moderate	Good	Good	Good
Freeze-thaw resistance	Good	Good	Low	Low	Low
Shrinkage	High	High	High	Low	High

The specific gravity of slag cement ranges from 2.85 to 2.95. The rough and angular-shaped ground slag in the presence of water and an activator, NaOH or $Ca(OH)_2$, both supplied by Portland cement, hydrates, and sets like Portland cement. Fly ash increases the workability of the fresh concrete because of its sphere particle shape. Fly ash extends the hydration process, allowing greater strength development and reduced porosity. Studies have shown that concrete containing more than 20% fly ash has a much smaller pore size distribution. Pozzolanic cement and slag cement have a lot in common in properties. It mainly reacts with the CH produced during the hydration of C_3S and C_2S. Table 3.9 summarizes the properties of blended cement and the details are discussed below:

(1) Workability: The workability of fly ash cement is better because of its spheri-

cal particles with dense surfaces.

(2) Hydration heat: Generally, the hydration of blended cement is slower than that of Portland cement or ordinary Portland cement, and therefore the early age strength and hydration heat are also lower (Figure 3.19).

(3) Strength: As shown in Figure 3.20, although the early age strength of slag and fly ash cement are lower, strength development after 28 days can exceed that of Portland cement. Usually, the higher the content of supplementary materials, the lower the early strength, but the higher the long-term strength. Therefore, it should be cured in a moist condition for a longer time.

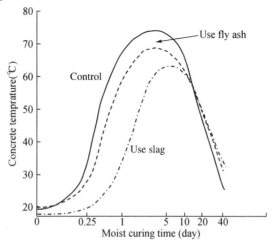

Figure 3.19　Concrete temperature of blended cement

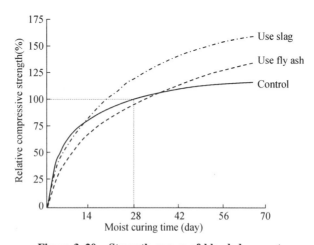

Figure 3.20　Strength curves of blended cement

(4) Impermeability: Blended cement has good impermeability and corrosion resistance because the secondary hydration generates more CSH gels and crystals filling the pores. Figure 3.21 shows that using fly ash and slag reduces both the accessible pore volume and the effective pore diameter. Generally, the permeability and absorption of hardened cement paste are reduced.

(5) Corrosion resistance: As shown in Figure 3.22, the volume expansion of blended cement during sulfate attack and alkali-aggregate reaction is also reduced. One reason is the improved impermeability of hardened cement paste. Another reason is CH and CAH are all consumed during the secondary hydration and therefore the risks of sulfate, salt, and acid corrosion are greatly reduced.

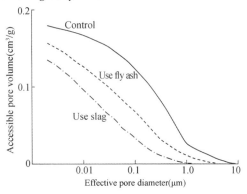

Figure 3.21   The relationship between pore volume and pore diameter

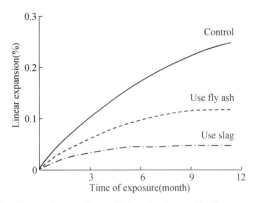

Figure 3.22   Expansion under sulfate attack or alkali-aggregate reaction

(6) Freeze-thaw resistance: Because blended cement has fewer micro air voids in hydrated cement mortar and can not provide space for the volume expansion of water during freezing. The freeze-thaw resistance of blended cement is lower than that of Portland cement.

(7) Shrinkage: Low content of supplementary materials usually have little effect on shrinkage. High content of slag may increase shrinkage, while fly ash tends to reduce the dry shrinkage. As the surface of fly ash is dense and spherical, its water absorption is low. The water content at the standard consistency of fly ash cement is low, and the dry shrinkage is also small.

Questions

1. What are reactions during the slaking and hardening of lime?
2. What are the primary chemical reactions during the hydration of Portland cement?
3. What are the four main chemical compounds in Portland cement?
4. Why gypsum is added to Portland cement?
5. What are the main hydration products of Portland cement?
6. Briefly discuss the initial setting and final setting of Portland cement.
7. Why is the early strength of slag cement lower while the long-term strength is higher than the same grade of ordinary cement?
8. What are the two main reactions of magnesium corrosion?
9. Discuss the measures to prevent cement corrosion.
10. A batch of 42.5 grade Portland cement has been stored in the factory for a very long time. The 28-day strength of samples is tested and the results are shown in the following table. Do the test results of this cement reach the requirements of the original strength grade? Can we determine the strength grade of the cement only by this test result?

Table 3.11 Strength test results

Ultimate compressive load (kN)	80.9	86.5	95.2	91.3	81.4	92.5
Ultimate flexural load (kN)	2.79		2.81		2.78	

4 Cement Concrete

Concrete is a mixture made of cementitious materials, water, coarse and fine aggregates, and admixtures. Since the invention of cement in 1824, cement concrete has been extensively used in buildings, pavements, bridges, tunnels, dams, etc. Cement concrete is easy to produce and can be cast into complicated shapes of structural members with good mechanical performance and durability. As summarized in Table 4.1, cement concrete has several key characteristics, making it a suitable material for civil engineering, as well as some drawbacks limiting its applications. This chapter covers the composition, workability, strength, deformation, durability, mix design, and admixtures of cement concrete.

Table 4.1 Advantages and disadvantages of cement concrete

Advantages	Disadvantages
• Concrete has high compressive strength ranging from 30 MPa to more than 100 MPa. • Concrete has good workability and can be cast into any desired shape. • Concrete has good durability and can last for centuries. • Concrete can be precast and then assembled on-site to improve construction efficiency and quality. • Steel can be embedded to make reinforced concrete and prestressed reinforced concrete which greatly improve the strength of concrete members. • Local ingredients including water and aggregates account for about 80% of the total weight of concrete which can save lots of costs. • Concrete has good resistance to heat and radiation.	• The tensile strength is low, only around 10% of its compressive strength. • Concrete is a brittle material and reinforcement is needed in many cases. • The weight of concrete is relatively high compared to its strength. • Concrete needs a long curing time during construction. • Concrete may have acid or salt corrosion problems.

4 Cement Concrete

4.1 Classification and Composition

4.1.1 Classification

There are many different types of concrete. The binders or cementing materials of concrete include inorganic binders such as cement and lime, organic binders such as asphalt and resin, and composites such as cement and polymer, and asphalt and cement. According to the density, concrete can be classified into four groups as below. The light weight concrete uses porous light weight aggregates such as expanded shale, clay, or slate materials to reduce the dead load on the structure and improve thermal insulation. The heavy weight concrete uses heavy weight aggregates such as slag to insulate radiation.

- Heavy weight concrete: >2500 kg/m^3
- Ordinary concrete: $1900-2500$ kg/m^3
- Light weight concrete: $1200-1900$ kg/m^3
- Super light weight concrete: <1200 kg/m^3

According to the strength, concrete can be classified into four groups as below. Superplasticizer is usually added to produce high strength concrete with a low water-cement ratio so as to reduce the cross-section of structural elements and self-weight of the structure.

- Super high strength concrete: >100 MPa
- High strength concrete: $60-100$ MPa
- Ordinary concrete: $20-60$ MPa
- Low strength concrete: <20 MPa

4.1.2 Composition

Concrete is a mixture of cement, aggregates, and water together with admixture that may be added to modify the placing and curing processes or the ultimate physical properties. Fresh concrete is a flowable plastic mixture and hardened concrete is a solid composite. As shown in Figure 4.1, the cement, water, and admixture which is optional are mixed to obtain a cement paste. Fine aggregates are added into the cement paste to obtain cement mortar and then coarse aggregates are added to produce cement concrete. There are several important mix proportion parameters for cement concrete:

(1) The water-cement ratio is the weight of water over the weight of cement. It is

Civil Engineering Materials

the most important factor for concrete, influencing both the strength and durability. The water-cement ratio usually ranges from 0.35 to 0.55. A higher water-cement ratio means higher porosity and lower strength.

(2) The paste-aggregate ratio is the weight of paste including cement and water over the weight of sand and coarse aggregates. It influences the workability. A higher paste-aggregate ratio usually means better workability since we have sufficient cement paste to coat and lubricate aggregates.

(3) The sand ratio is the weight of sand over the sum of the weight of sand and coarse aggregates. It influences the workability of concrete. When the sand ratio is low, the concrete mainly contains coarse aggregates and the workability is poor. However, when the sand ratio is too high, the concrete does not have enough cement paste to coat the fine and coarse aggregates and the workability is also poor.

Figure 4.1 Composition of cement concrete

Figure 4.2 shows the relative volume proportion of each ingredient in different concrete. Cement accounts for 7% to 15% of the total. Water accounts for 14% to 21%. Sand accounts for 24% to 30%. Coarse aggregates account for 31 to 51%. Concrete

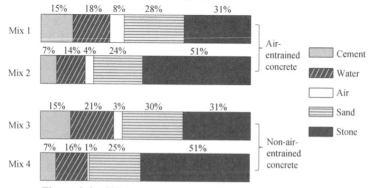

Figure 4.2 Volume proportion of ingredients in concrete

also contains a specific amount of air voids. Air-entrained concrete increases air content and can improve the resistance of concrete to freeze-thaw cycling because air voids in concrete provide space for the volume expansion of freezing water.

1) Cement

Cement is an essential part of concrete as a binding or cementing agent. It forms a paste with water and progressively sets and hardens in the air or water. Its content usually ranges from 240 to 400 kg/m^3. Cement is obtained by burning a mixture of limestone and clay, with appropriate proportions, in rotating kilns with small quantities of other minerals, thereby obtaining the clinker which is then milled and mixed with gypsum. The quality of cement is critical for the strength and durability of concrete.

2) Water

Water mainly influences the hydration and workability of cement concrete. Its content usually ranges from 140 to 230 kg/m^3. The amount of water influences the strength, durability, and workability of concrete. Only one-third of the water hydrates. The rest water evaporates or becomes capillary or interlayer water. The quality of the water should also be assessed. Water may contain impurities that may interfere with the setting of the cement, influence the concrete's strength and form stains on the surface, and contain chlorides that cause the corrosion of reinforcing steel.

3) Aggregates

Aggregates are the granular materials used in concrete, ranging from 1600 to 1800 kg/m^3, or from 60% to 80% by volume. The functions of aggregates in concrete include forming a strong skeleton, acting as a filler, reducing costs, providing strength, durability, and abrasion resistance, and reducing changes in volume due to the hydration and drying processes. The mineral characteristics of aggregates influence the properties of hardened concrete, while its geometric characteristics influence the workability of fresh concrete. The quality of aggregates should be tested especially for the alkali activity.

4) Admixtures

Admixtures are additional materials that may be added to the concrete in small quantities to improve the properties of fresh and hardened concrete, including increasing flowability strength, and air content, and reducing water content, shrinkage, curing time and costs. Using admixtures has been very common in concrete. However, the poor mix design can not be improved only by using admixtures.

4.2 Workability of Fresh Concrete

4.2.1 Workability

Workability describes how easily fresh concrete can be mixed, placed, consolidated, and finished while maintaining its homogeneity throughout these operations. The workability of fresh concrete mix is mainly evaluated by different tests including flowability, cohesiveness, and water retention.

1) Flowability

Figure 4.3 shows the flowability of fresh concrete. Flowability is also called consistency, fluidity, or plasticity. It measures the ability of concrete mix to flow and fill every corner of the mold evenly and densely, under the influence of self-weight or mechanical vibration. Poor flowability or consistency will leave large voids, causing defects in concrete. It is noted that the flowability of concrete is related to the equipment and structure. For example, roller-compacted concrete requires low flowability while casting thin columns and beams needs high flowability.

Figure 4.3　Flowability of fresh concrete

2) Cohesiveness

The concrete should remain as a homogeneous uniform mass throughout with no segregation. Segregation is the separation of fresh concrete ingredients from each other, resulting in the non-uniform mix, in which cement paste comes to the top and aggregates settle at the bottom. As shown in Figure 4.4, segregation can cause water-rich pockets under aggregate particles and water film on the surface. Water-rich pockets are weak zones in concrete while surface water may cause surface defects and shrinkage cracking. Segregation influences the strength and durability of concrete, and can be caused by poor mix design, and over vibration or compaction.

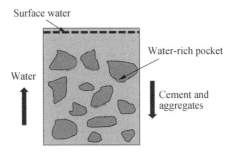

Figure 4.4　Segregation in fresh concrete

3) Water retention

Water retention means fresh concrete should retain water and has no bleeding. Bleeding in fresh concrete is the free water in the mix migrates to the surface and forms a paste of cement. As shown in Figure 4.5, the water on the surface evaporates to form the typical surface plastic micro-cracks after hardening. Segregation and bleeding may occur simultaneously. Figure 4.6 shows the bleeding in fresh concrete with reinforced bars, after placing, the aggregates tend to go down while free water tends to go upward. For poorly mixed concrete, after several hours, the water gathers on the lower surface of the reinforcement, creating water-rich pockets and leaving defects in the concrete. Surface water evaporates to form the typical surface plastic micro-cracks after hardening.

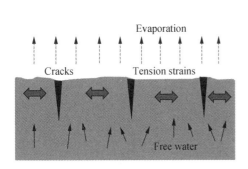

Figure 4.5　Bleeding in fresh concrete

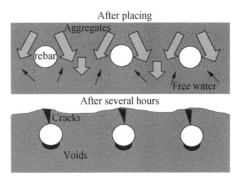

Figure 4.6　Bleeding in fresh concrete with reinforcement

4) Slump test

As shown in Figure 4.7, the most common method for characterizing concrete flowability is the slump test. It is easy to perform and also appropriate for most concrete. According to specifications GB/T 50080—2016 and JTG 3420—2020 T0522, during the test, the cone is placed on the flat plate, and then concrete is filled into the slump cone in three layers. The height of each layer is about 1/3 of the height of the cone after tamping. Each layer should be tamped uniformly 25 times using the rounded end steel rod. After filling the cone, the concrete is struck off flush with the upper edge of the slump cone. Then the cone is carefully lifted vertically upwards without lateral or torsional motion, and the cone and the mix are discharged on the flat plate. The height difference between the cylinder height and the highest point of the concrete sample after the collapse is measured, which is the slump of the fresh concrete mix, in mm. The slump of the concrete is measured by determining the vertical difference between the top of the mold and the slumped concrete. The greater the slump, the greater the fluidity of the concrete mix. The flowability of concrete mix can be classified into four levels, i.e. low plastic, plastic, flowable, and high flowable depending on its slump as shown in Table 4.2.

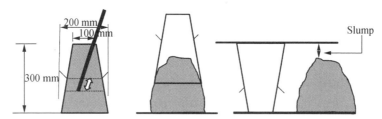

Figure 4.7 Slump test

Table 4.2 Classification of concrete slump

Level	Types	Slump (mm)
T1	Low plastic	10 – 40
T2	Plastic	50 – 90
T3	Flowable	100 – 150
T4	High flowable	≥160

5) Vebe consistency test

For concrete with low flowability, the slump test can not differentiate concrete with different consistency or plasticity, and the Vebe consistency test is recommended according to specifications GB/T 50080—2016 and JTG 3420—2020 T0523. The Vebe consistency test involves placing the slump cone in a cylinder installed on a vibrating table. The fresh concrete is filled into the slump cone which is then lifted, and a transparent disc is placed on top of the fresh concrete. The vibrating table is started and the time taken for the transparent disc to be fully in contact with the concrete is measured as the Vebe time, as shown in Figure 4.8.

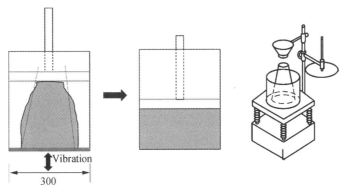

Figure 4.8　Vebe consistency test

The relationship between Vebe consistency and slump is shown in Figure 4.9. It can be seen that the slump test can not differentiate concrete mix with low flowability, for which the Vebe consistency test is needed. According to specification GB/T 50080—2016, the flowability of concrete mix can be classified into four levels based on the Vebe consistency, as shown in Table 4.3.

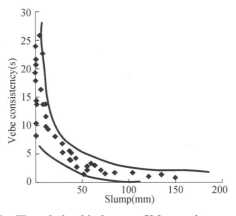

Figure 4.9　The relationship between Vebe consistency and slump

Table 4.3 Levels of concrete flowability based on Vebe consistency

Level	Types	Vebe time (s)
T1	Super dry	≥31
T2	Very dry	30 – 21
T3	dry	20 – 11
T4	Semi-dry	10 – 5

4.2.2 Influencing Factors

Concrete's workability is influenced by internal factors including material properties and proportions, as well as external factors including environment, time, and handling. The influencing material properties include cement and aggregates. The key material proportions are water content, water-cement ratio, paste-aggregate ratio, sand ratio, and admixtures.

1) Cement

The type, fineness, mineral composition, and particle shapes of cement influence the required water content to achieve the standard consistency and workability. Generally, fly ash cement has high flowability because of the spherical particle shape of the fly ash. Properly increasing the fineness of cement can improve the cohesion and water retention of concrete mix, and reduce bleeding and segregation.

2) Aggregates

The maximum particle size, shape, surface texture, gradation, and water absorption of aggregates all influence the workability of fresh concrete. Concrete made with gravel has higher flowability than those made with crushed aggregates. The rounder the aggregate particle, the less energy is needed to let the aggregate move around within the concrete. Good gradation improves workability. Increasing the maximum particle size can reduce the total surface area of aggregates and therefore improve the workability of the mixture.

3) Water content

The most important factor is water content since water is the most fluid constituent. It increases the distance between particles, which makes it easier for the mixture to move. The flowability of fresh concrete mainly depends on a layer of water film absorbed on the surface of aggregates and cement particles to make the particles more lubricated. If the water content is too low, the water film will be thin and the lubrication effect is poor. However, if the water content is too high, the cohesiveness of concrete is very poor, which may cause segregation or bleeding. When maintaining the same aggregates and a 0.4 - 0.8 water-cement ratio, the slump of concrete is mainly determined by water content.

4) Water-cement ratio

The consistency of cement paste is mainly determined by the water-cement ratio. If the water-cement ratio is too low, it will cause very low flowability, making concrete difficult to handle. If the water-cement ratio is too high, the flowability of concrete is high. But it will make concrete susceptible to segregation and bleeding. In addition, a high water-cement ratio also reduces the strength and durability of hardened concrete. During mix design, the water-cement ratio is mainly determined based on strength requirements. To adjust the flowability of concrete, it is usual to maintain the same water-cement ratio and change the amount of water and cement simultaneously, which is actually to change the paste-aggregate ratio.

5) Paste-aggregate ratio

Cement paste coats aggregate particles to reduce the friction between aggregate particles and provides the flowability of concrete. For the same water-cement ratio, the higher the paste-aggregate ratio, the higher the flowability of the concrete. However, if the paste-aggregate ratio is too high, the cohesiveness and water retention of concrete are compromised. The strength and durability of hardened concrete might be low and the cost of concrete significantly increases. Usually, for the mix proportions to meet the requirements of workability, strength, and durability, a low paste-aggregate ratio is preferred.

6) Sand ratio

The change in sand ratio leads to the change in the specific surface area of aggregates and the amount of cement paste to coat aggregates and therefore the workability of concrete is changed. Figure 4.10 shows the relationship between the slump and sand ratio of concrete. A high sand ratio means a high specific area of coarse and fine aggregates and more cement paste is needed to coat aggregates. A low sand ratio means the amount of cement mortar is reduced, which can not effectively coat coarse aggregate particles to provide sufficient flowability and may cause segregation. Therefore, with the increase of sand ratio, the slump increases first and then decreases. The optimum sand ratio can maximize flowability while maintaining good cohesiveness and water retention.

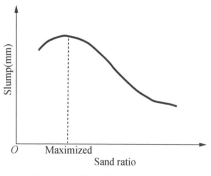
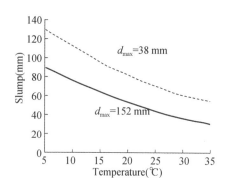

Figure 4.10 The relationship between sand ratio and slump

Figure 4.11 The relationship between temperature and slump

7) External factors

Time, temperature, humidity, and wind speed are the main external and environmental factors that influence the workability of concrete. After mixing, the workability of concrete gradually decreases with the increase of time. The decrease in concrete workability is mainly determined by the hydration rate of cement and the evaporation rate of water. On one hand, the hydration of cement consumes water; on the other hand, the hydration products thicken and further hinder the sliding between particles. As shown in Figure 4.11, the workability of concrete decreases as the ambient temperature increases due to the increased water loss caused by fast evaporation and cement hydration. Similarly, wind speed and humidity also influence workability. It can also be seen that the large-size aggregate reduces workability.

4.2.3 Measures to Improve Workability

According to specification JGJ 55—2011, the slump of cement concrete is determined based on the construction method (manually or with equipment), structural conditions (section size and reinforcement), and experience. Generally, a higher slump is selected as the casting difficulty increases. For the structure with a small section size, complex shape, or heavy reinforcement, a large slump should be selected. For the structure with less reinforcement or easy to cast, a small slump is selected to save the amount of cement paste or cement and to improve strength.

  Measures to improve concrete workability mainly include selecting proper materials and mix proportion, using admixtures, and improving construction technologies such as using proper casting and handling tools, while maintaining needed strength, durability, and costs.

(1) Selecting proper materials and mix proportion: Selecting proper materials includes using sand and coarse aggregates with good gradation, and using coarser gradation to reduce surface area. It is recommended to use the mix proportion with a small sand ratio to improve concrete quality and save the amount of cement. It is also recommended to increase the paste-aggregate ratio when the slump of concrete is too low, and decrease the paste-aggregate ratio when the slump is too high.

(2) Using admixtures: Using admixtures is a very common and important measure to adjust the performance of concrete, commonly using water reducer, superplasticizer, pumping agent, etc. Admixtures improve the workability of fresh concrete, the strength of concrete, and the durability of concrete, while reduce the amount of cement. Using a retarder can reduce the workability during long-distance transportation.

(3) Improving the construction technologies: A high-power concrete mixer can improve the lubrication efficiency of water, and the high-efficiency vibrating equipment can also obtain better compactness. Fast construction speed can also help ensure sufficient workability during casting.

4.3 Properties of Hardened Concrete

4.3.1 Mesoscale Structure

As shown in Figure 4.12, the mesoscale structure of hardened cement concrete includes three parts: aggregates, hardened cement paste (HCP), and interfacial transition zone (ITZ). Cracking tends to propagate through the ITZ and the weak zone in HCP in concrete. The HCP contains hydration products, cement residues, free water, voids, etc. As shown in Figure 4.13, the ITZ is a weak interface between aggregates and bulk cement paste, which is usually 30 – 50 μm thick and contains a large amount of CH and CASH. The bulk cement paste contains a great number of compacted CSH gel crystals.

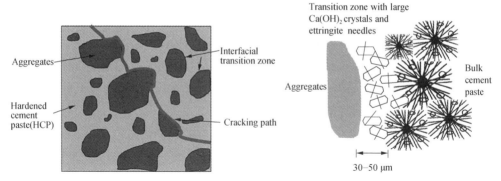

Figure 4.12 Mesoscale structure of hardened cement concrete

Figure 4.13 Composition of interfacial transition zone

The ITZ has high porosity, loose structure, low density, and low strength. One reason is the high local water-cement ratio because of the water film on the surface of aggregates. Another reason is the "Wall effect", whereby the cement grains cannot pack as efficiently next to the aggregate surface as they can in the bulk paste. As shown in Figure 4.14, there are more cement particles on a cross-sectional surface in the bulk cement paste than on the surface of aggregates.

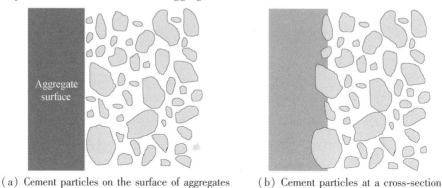

(a) Cement particles on the surface of aggregates   (b) Cement particles at a cross-section

Figure 4.14 "Wall effect" of ITZ

4.3.2 Strength of Concrete

1) Compressive failure

In different structural members including foundations, beams, and columns, concrete is mostly used for compressive loads. Figure 4.15 shows the typical load-deformation curve and the failure process of concrete under compressive loads, which can be divided into four stages. The original concrete specimen is a composite of aggregates, HCP, ITZ, micro-cracks, and voids.

(1) Stage I is from 0 to 30% of the ultimate load, during which the micro-cracks at ITZ do not change significantly. The load-deformation relationship is linear.

(2) Stage II is from 30% to 70% of the ultimate load, during which the amount, width, and length of micro-cracks at ITZ become larger. The load-deformation relationship is no longer linear.

(3) Stage III is from 70 to 100% of the ultimate load, during which the micro-cracks at ITZ propagate into the bulk cement paste and connect rapidly. The deformation increase rapidly and the load reaches the ultimate maximum load.

(4) Stage IV is after 100% of the ultimate load, during which more cracks connect and the load reduces till the fracture of the specimen.

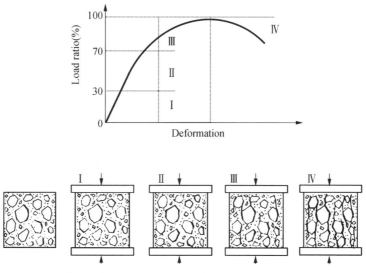

Figure 4.15　Four stages of compressive failure

2) Compressive strength

The compressive strength is the most important structural design parameter for concrete and is used to characterize concrete grade. The specimens are tested by applying axial compressive loads with a specified rate of loading until failure using a high-capacity hydraulic compression test machine. It is the maximum load supported by the specimen divided by the cross-sectional area. According to specifications GB/T 50081—2019 and JTG 3420—2020 T0553, 150 mm × 150 mm × 150 mm standard size cubic specimens are prepared, cured in the standard condition (20 ℃ and ≥90 % humidity) and then tested at 28th day to obtain the cubic compressive strength, denoted as f_{cu}. No less than three specimens should be tested and the average value is used as the cubic compressive strength. The 100 mm or 200 mm cubic specimens can be used for concrete with small or large maximum aggregate size. The test results of 100mm and 200mm cubic specimens should time 0.95 and 1.05 respectively to obtain the compressive strength at standard size. Generally, large specimens have less variability and are close to the true value of the strength.

Structure designing engineers can not directly use the cubic compressive strength, because it is an average value meaning 50 % of the produced concrete have strength lower than this value. The designer needs to use strength lower than the average of our test results to ensure reliability. The characteristic compressive strength is defined as the strength of the concrete below which not more than 5 % of the test results are expected to fall (Figure 4.16). According to the normal distribution of the test results, the required compressive strength $f_{cu,k}$ is the average compressive strength minus 1.645 times the standard deviation as shown in Equation (4.1).

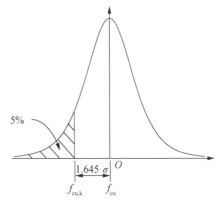

Figure 4.16 Distribution of the compressive strength test results

$$f_{cu,k} = f_{cu} - 1.645\sigma \qquad (4.1)$$

where $f_{cu,k}$ = characteristic compressive strength (MPa);

f_{cu} = average cubic compressive strength (MPa).

Characteristic compressive strength can be used to represent concrete strength grade, recoded as "C" + "Characteristic compressive strength in MPa". Specification GB/T 50107—2010 defines a total of 12 grades of concrete, ranging from 7.5 to 60 MPa and including C7.5, C10, C15, C20, C25, C30, C35, C40, C45, C50, C55, and C60. The most commonly used strength grades are C30 to C40. C7.5 and C15 are usually used for foundations or structures with very small loads. Pre-stressed structure members require C30 or above grades.

3) Axial compressive strength

To design concrete columns, the axial compressive strength f_{cp}, tested with column or cylindrical specimens with a high height over width ratio, should be used. As shown in Figure 4.17, during the compressive strength test, the specimen has lateral deformation which can be calculated by Poisson's ratio. The loading heads have horizontal friction forces on the surface of the two ends. The cracks near the two ends of the specimen are at approximately 45° to the axis due to the effects of the vertical compression and horizontal friction. The cracks away from the ends of the specimen are parallel to the axis. The surfacing friction of the loading head has a confining constraint on the bottom and top of the specimen and tends to increase the compressive strength. For columns, failure under compressive loads is caused by the propagation of vertical cracks, therefore lateral constraint can greatly increase compressive strength.

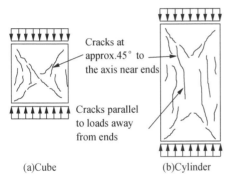

Figure 4.17 Compressive strength test in cube and column

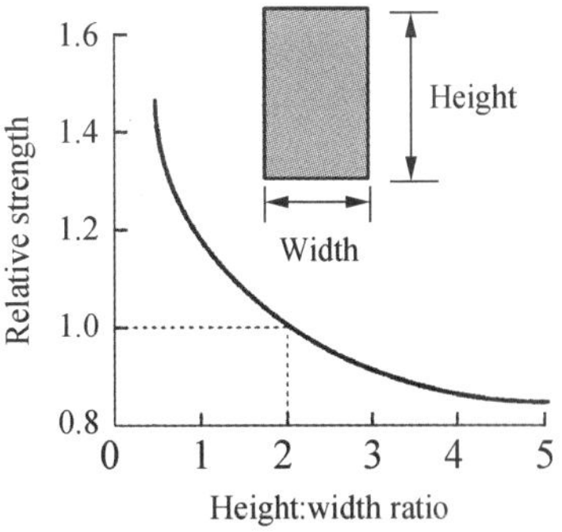

Figure 4.18 Stress-strength ratio

As shown in Figure 4.18, the higher the height-width ratio, the lower the tested relative strength. When the height-width ratio is low, the friction confinement at ends becomes significant and the tested compressive strength is relatively higher. When the height-width ratio is high, the tested compressive strength is less influenced by the friction confinement at the ends. However, if the height-width ratio is too high, the specimen is susceptible to stability failure and also has tested compressive strength. In the specifications GB/T 50081—2019 and JTG 3420—2020 T0555, the axial compressive strength test uses the 150 mm × 150 mm × 300 mm column specimen. For the C10-C55 concrete, the axial compressive strength is around 70 % to 80 % of the cubic compressive strength. The specifications ASTM C39/C39M-21 and JTG 3420—2020 T0554 recommend using a cylindrical specimen with a diameter of 150 mm and a height of 300 mm for the concrete compressive strength test.

Loading rate also has a significant influence on the tested strength. For a large loading rate, cracks do not have enough time to propagate before fracture, and the tested strength is higher. For a small loading rate, cracks will have enough time to propagate and the tested strength is lower. As shown in Figure 4.19, the strain or deformation increases significantly for a small loading rate indicating the propagation of cracks and the accumulation of damage. The smaller the loading rate, the lower the ultimate strength. When the force reaches 80% of the ultimate load, cracks will propagate till fracture even if the load is maintained constant.

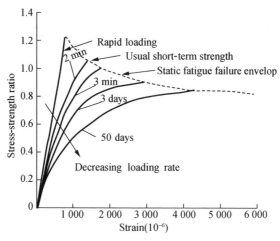

Figure 4.19 Influence of loading rate on the compressive strength and deformation

4) Splitting tensile strength

The tensile strength of concrete is only 5% to 10% of its compressive strength and concrete is not designed to bear the tensile force. But the tensile strength is meant to estimate the cracking due to shrinkage and temperature change. Concrete pavement slabs require sufficient flexural strength. The direct tensile strength of concrete is very difficult to perform due to the eccentricity effects, and the splitting tensile strength or the indirect tensile strength is usually tested instead. In the specifications GB/T 50081—2019 and JTG 3420—2020 T0560, the splitting tensile strength is to apply a line force on the top and bottom of the 150 mm × 150 mm × 150 mm cubic specimen with a steel bar. The force has a splitting tensile effect at the vertical axis of the specimen. The fracture of the specimen occurs along its vertical axis, due to tension developed in the transverse direction as shown in Figure 4.20. The splitting tensile strength f_{ts} can be calculated by Equations (4.2) and (4.3).

$$f_{ts} = \frac{2F}{\pi A} = 0.637 \frac{F}{A} \tag{4.2}$$

where F = load at failure (N);
A = cross-sectional area (mm²).

The specification ASTM C496/C496M-11 uses a cylindrical specimen of 150 mm in diameter and 300 mm in height and the force is applied on the lateral side of the specimen. The splitting tensile strength f_{ts} is calculated as below.

$$f_{ts} = \frac{2P}{\pi L d} \tag{4.3}$$

where F = load at failure (N);
L = length of the specimen (mm);
d = diameter of the specimen (mm).

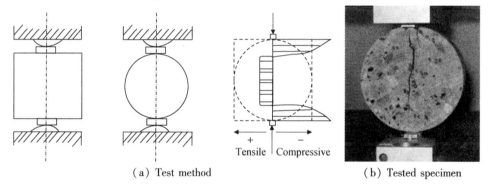

(a) Test method (b) Tested specimen

Figure 4.20 Splitting tensile strength test

5) Flexural strength

The flexure strength is important for the design and construction of concrete pavement. According to specifications GB/T 50081—2019 and JTG 3420—2020 T0558, the flexure strength test uses a 150 mm × 150 mm × 550 mm beam specimen and is performed with a four-point bending test, as shown in Figure 4.21. The four-point loading ensures a constant bending moment without any shear force applied on the middle third of the specimen. If fracture initiates in the tension surface within the middle third of the span, the flexure strength f_{cf}, also called modulus of rupture, is calculated as below.

$$f_{cf} = \frac{PL}{bh^2} \tag{4.4}$$

where P = load at failure (N);

L = span length, 450 mm;

b = width of the specimen, 150 mm;

h = depth of the specimen, 150 mm.

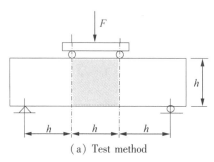
(a) Test method

(b) Tested specimen

Figure 4.21 Flexural strength test

6) Rebound hammer test

In addition to those destructive strength tests, several non-destructive tests can be used to estimate the strength or other mechanical properties of concrete, especially for the in-situ concrete structure members. Figure 4.22 shows a Schmidt rebound hammer. The rebound hammer test is a non-destructive testing method of concrete using a Schmidt rebound hammer to test the hardness of concrete. The amount of rebound is read on a scale attached to the device to obtain the hardness and estimate the compressive strength of concrete.

Figure 4.22 Schmidt rebound hammer

4.3.3 Strength Influencing Factors

Similar to factors influencing concrete workability, factors influencing concrete strength also include internal causes such as material properties and proportions, and external factors such as curing time, humidity, and temperature.

1) Cement

The HCP plays a vital role in the strength and durability of cement concrete. The strength and quality of HCP are mainly determined by the cement and water-cement ratio. For a fixed material proportion, the higher the cement strength grade, the higher the concrete strength.

2) Aggregates

Aggregate particle shape, angularity, and surface texture also influence the strength of concrete. Generally, cubic shape, high angularity, rough texture, and more fractured surfaces can help improve the strength of concrete.

3) Water-cement ratio

Only one-third of the water in cement concrete hydrates. The excess water is to improve obtain flowability, which remains in the concrete after hardening and becomes capillary water. Generally, for a fixed cement content, the higher the water-cement ratio, the lower the concrete strength. However, when the water-cement ratio is too low, the concrete is hard to handle and may cause poor casting quality, reducing the strength of the concrete. As shown in Equation (4.5), the concrete strength can be expressed as a function of cement strength, the water-cement ratio, and the aggregate type.

$$f_{cu} = A f_{ce} \left(\frac{C}{W} - B \right) \quad (4.5)$$

where f_{cu} = 28-day compressive strength of concrete (MPa);

f_{ce} = 28-day compressive strength of cement;

C/W = cement over water ratio;

A and B = coefficients related to the type of coarse aggregates as shown in Table 4.4.

Usually, when the water-cement ratio is higher than 0.65, the difference between gravel and crushed aggregates is not significant; when the water-cement ratio is lower than 0.4, the strength of concrete using crushed aggregates is 30% higher than that of gravel.

Table 4.4 Coefficient for aggregate types

Types	A	B
Crushed aggregates	0.53	0.20
Gravel	0.49	0.13

4) Paste-aggregate ratio

The paste-aggregate ratio has a significant influence on both the workability and strength of concrete. For a fixed water-cement ratio, if the paste-aggregate ratio is too low, the cement paste can not coat aggregates, leaving large air voids in concrete, reducing the strength. If the paste over aggregates ratio is too high, the skeleton of coarse aggregates is weak. There is also too much water causing too many pores in concrete. In addition, too much paste increases shrinkage, which may cause micro-cracks in concrete. Those all reduce the strength of concrete. Therefore, as the increase of paste-aggregate ratio, the strength of concrete increase and then decrease.

5) **Time**

The strength of concrete increases as the curing age increases. The strength increases rapidly in the first 28 days and then the speed slows down. Therefore the 28-day strength is usually used to characterize the strength of concrete. Figure 4.23 and Equation (4.6) show the relationship between compressive strength and the curing age in the first 28 days, based on which we can either estimate the 28-day strength based on early-age strength or estimate early-age strength based on the 28-day strength to determine the time of demolding.

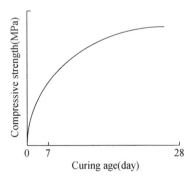

Figure 4.23 Relationship between concrete Compressive strength and curing age

$$f_n = \frac{f_{28}}{\lg 28} \lg n \qquad (4.6)$$

where f_{28} = 28-day compressive strength (MPa);

f_n = compressive strength at n days (MPa), $n > 3$.

6) **Moisture**

Moisture is critical for the strength growth of concrete. Moisture-curing time significantly improves strength. Insufficient moisture will leave more unhydrated cement and therefore the strength is lower. As shown in Figure 4.24, the longer the moisture-curing time, the higher the long-term strength. The concrete strength keeps increasing for one year when it is moist cured. Synthetic concrete curing blankets are usually used to preserve moisture for concrete at the job site. For ordinary cement, moisture-curing time should be no less than 7 days. For slag and fly ash cement, longer moisture-curing time (>14 days) is recommended. For the concrete with impermeability requirements, moisture-curing time should be no less than 14 days.

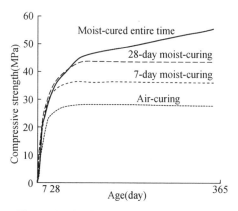

Figure 4.24 Strength of concrete cured at different moisture conditions

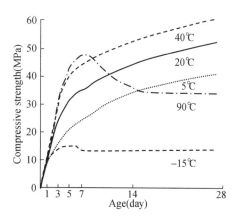

Figure 4.25 Strength of concrete cured at different temperatures

7) Temperature

The temperature has a significant effect on the early strength of concrete. Figure 4.25 shows the strength of concrete cured at different temperatures. Generally, in the range of 4 – 40 ℃, the increase of curing temperature improves the hydration and hardening of cement and the early strength of cement. However, if the early curing temperature is too high (>40 ℃), the hydration rate of cement will accelerate, and a large number of hydrates will form around the unhydrated cement particles, which will hinder the further hydration of cement. Moreover, these rapidly formed hydration products have coarse crystals and high porosity, which is unfavorable to the later strength growth of concrete. When the temperature drops below 0 ℃, the hydration stops, and the volume expansion (9%) of freezing water causes severe damage to the internal structure of the concrete. Therefore, special attention should be paid to heat preservation and maintenance in winter construction.

Based on the above discussion, it is easy to find some measures to improve the strength of concrete. Using high-strength cement is a direct way of improving strength. Reducing water-cement ratio can reduce porosity and form a denser structure. Using admixtures such as water reducers will also reduce the water-cement ratio. Keeping sufficient high humidity and temperature during curing can improve strength. We can also use mixing and handling equipment to improve the consolidation and construction quality.

4.3.4 Stress-strain Relationship

The deformation of concrete includes load-independent deformations such as shrinkage and load-dependent deformations which can be further classified into short-term and long-term deformations.

1) Short term deformation

The short-term deformation behavior of concrete is under a high loading rate during tests. Figure 4.26 shows the stress-strain relationship of aggregates, HCP, and concrete during a compressive strength test. Since the strength of aggregates is much higher than that of the traditional HCP, aggregates exhibit elastic behavior. HCP exhibits a near elastic behavior for lower loads and a plastic behavior after the elastic limit. The strength of concrete is mainly determined by the strength of HCP and is usually lower than HCP due to the ITZ. Because of the plastic behavior of concrete, during the high-level cyclic loading as shown in Figure 4.27, concrete accumulates plastic deformations, indicating the damage caused by the cyclic loading.

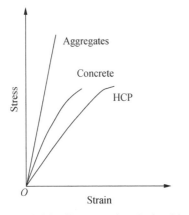
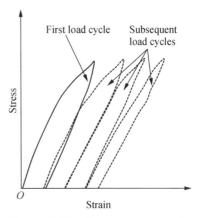

Figure 4.26 Stress-strain relationship of aggregates, HCP, and concrete

Figure 4.27 Concrete stress-strain relationship under cyclic loading

As shown in Figure 4.28, from the compressive strength test, different elastic moduli can be calculated. The tangent modulus is the slope of a line tangent to the stress-strain curve at a point of interest. The initial tangent modulus is the tangent modulus at the original point. The secant modulus, which is the secant modulus from the original point, is the slope of a line drawn from the starting point of interest of the stress-strain diagram and intersecting the curve at the endpoint of interest. The initial secant modulus is often used as the modulus for structural design. In the specification GB/T 50081—2019, the modulus is calculated using the strain at 0.5 MPa stress and 30% of the ultimate strength. The Poisson's ratio is the relationship between lateral strain and axial strain under simple axial stress in the elastic stage. The Poisson's ratio of concrete ranges from 0.15 to 0.20. Usually, the higher strength, the lower the Poisson's ratio.

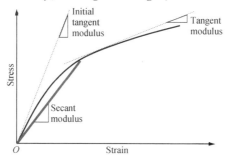

Figure 4.28 Definition of modulus of concrete

2) Long-term deformation

Constant loading such as the self-weight of the structure is a very common occurrence causing long-term deformation of concrete. As shown in Figure 4.29(a), creep means when subjected to constant stress, the strain within the specimen will gradually increase. Under constant loading, concrete has immediate elastic deformation at first, and then the viscous deformation and plastic deformation gradually increase. When the stress

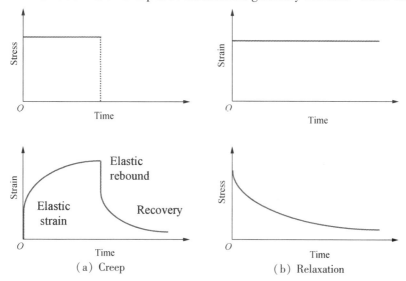

Figure 4.29 Definition of creep and relaxation

is unloaded, elastic strain recovers immediately, viscous deformation recovers slowly, and the plastic deformation retains.

As shown in Figure 4.29(b), relaxation means when subjected to a constant deformation or strain, the stress within the concrete gradually dissipates due to relaxation. Relaxation is the decrease in stress under constant strain, while creep is the increase in strain under constant stress.

4.3.5 Shrinkage

Shrinkage of concrete is the decrease in length or volume of concrete resulting from changes in moisture content or chemical composition. Shrinkage of concrete can be caused by many reasons including water evaporation, chemical shrinkage, autogenous shrinkage, drying, carbonation, and temperature change.

1) Plastic shrinkage

Plastic shrinkage is the slight shrinkage (1 ‰) of fresh cement paste due to the loss of water. Shrinkage cracks are the surface micro-cracks due to the evaporation of water, especially when bleeding occurs (Figure 4.30). They are mostly short and irregular cracks. The water in fresh concrete goes upward and evaporates, causing tensile strains on the surface of concrete and inducing plastic cracks.

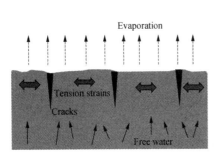

Figure 4.30　Plastic shrinkage at the surface of concrete

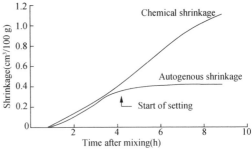

Figure 4.31　Autogenous shrinkage and chemical shrinkage

2) Autogenous and chemical shrinkage

The shrinkage of fresh cement paste during hydration includes chemical shrinkage and autogenous shrinkage. At low water-cement ratio (<0.42), water is rapidly drawn into the hydration process and the demand for more water creates very fine capillaries. The shrinkage caused by the surface tension within the capillaries before the setting of cement is called the autogenous shrinkage. Chemical shrinkage occurs because the volume of hydration products is smaller than that of the water and cement after setting. As shown in Figure 4.31, the autogenous shrinkage stops after setting, while the chemical shrinkage continues to increase as the process of hydration goes.

3) Drying shrinkage

Drying shrinkage occurs during and after hydration. It is caused by the evaporation of capillary water, which also occurs in hardened concrete. 15%–30% drying shrinkage occurs in the first two weeks, and 65%–85% drying shrinkage occurs in the first year. The hardened concrete shrinks and expands due to the change in humidity of the environment. As shown in Figure 4.32, the maximum shrinkage occurs in the first drying and a part of this shrinkage is irreversible. Further drying and wetting cycles result in reversible shrinkage. Factors influencing drying shrinkage include the volume of the structure, cement properties, water-cement ratio, curing, etc.

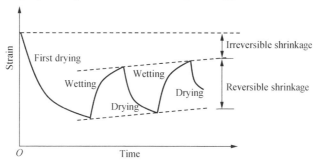

Figure 4.32 Drying shrinkage

4) Carbonation shrinkage

Carbonation shrinkage is that when the concrete is exposed to air, calcium hydroxide reacts with carbon dioxide to form calcium carbonate deposition, which is less in volume than the original calcium hydroxide, causing shrinkage. As shown in Figure 4.33, drying shrinkage decreases with the increase in humidity. Carbonation shrinkage is maximized at around 50% humidity since the carbonation reaction needs both moisture and air. Therefore, the total shrinkage is also maximized at around 50% humidity.

5) Thermal expansion

Thermal expansion has a significant influence on the concrete structure design. The thermal expansion coefficient of concrete is around 1×10^{-5} $\mu\varepsilon/℃$, which is nearly the same as that of steel. This extremely fortuitous coincidence is the key to the steel-reinforced concrete, meaning that when they are subject to heat (or cold), they expand (or shrink) at essentially the same rate. As shown in Figure 4.34, the thermal expansion coefficient of HCP and concrete is higher than that of aggregates and achieves the highest value at around 65 % humidity.

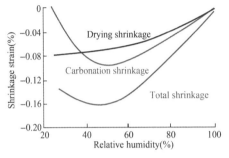

Figure 4.33 Carbonation shrinkage   Figure 4.34 Thermal expansion

4.3.6 Durability of Concrete

The durability of concrete is the ability to resist weathering action, chemical attack, abrasion, or any other process of deterioration, and retain its original form, quality, and serviceability when exposed to its environment. The durability of concrete is determined by the quality of cement, aggregates, water-cement ratio, and material proportions. Good durability needs proper design, proportioning, placement, finishing, testing, inspection, and curing.

1) Abrasion

The abrasion resistance of concrete is a very important property of concrete pavement as it is subjected to the loading of traveling wheels. Abrasion resistance depends closely on the abrasion resistance of its constituent coarse aggregates and cement paste. The specification JTG 3420—2020 T0567 provides a test method for determining the abrasion resistance of either concrete or mortar surface using the rotating-cutter method and the weight loss is used to measure concrete abrasion resistance.

2) Impermeability

Impermeability is the ability of concrete to resist penetration by water or other liquids, gas, ions, etc. Impermeability is critical for durability because it controls the rate of entry of moisture that contains aggressive chemicals and the movement of water during heating and freezing. Impermeability is determined by the interconnect voids and is different from porosity. As shown in Figure 4.35, the upper concrete has high porosity, however it has low interconnected channels and therefore its permeability is low; while the bottom concrete has high interconnect channels although its porosity is low, and therefore the permeability is high. Figure 4.36 shows a test system for measuring concrete permeability under steady-state flow. Specifications GB/T 50082—2009 and JTG 3420—2020 T0568 provide the chloride permeability test to check the concrete's ability to resist chloride ion penetration.

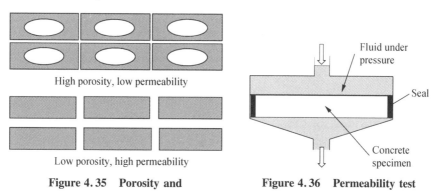

Figure 4.35 Porosity and permeability in cement concrete

Figure 4.36 Permeability test

3) Freeze-thaw resistance

Freeze-thaw damage is caused by the volume expansion of freezing water in concrete. The volume increase of frozen water creates hydraulic pressure, expelling water through unfrozen pores to a free space and causing micro cracking. As shown in Figure 4.37, the specification JTG 3420—2020 T0665 provides a rapid freeze-thaw test method to evaluate the freeze-thaw resistance of concrete beam specimens. The relative dynamic modulus (RDM) or the resonant frequency, weight loss, and change of length are used to evaluate the freeze-thaw resistance.

Figure 4.37 Concrete rapid freeze-thaw test

4) Corrosion

Similar to the corrosion discussed in the chapter on cement, corrosion of concrete includes different physical reactions and chemical reactions such as dissolving, ion exchange, and forming expansive components. Soft water corrosion is the dissolving of hydration products, especially calcium hydroxide, in soft water. The reactions between salt or acids and calcium hydroxide, can either form calcium dichloride which is soluble and may cause steel rusting, or form gypsum which can react with calcium aluminate hydrates to form calcium sulfoaluminate hydrates causing volume expansion and cracking. Figure 4.38 shows the steel rusting and cracking due to expansion.

   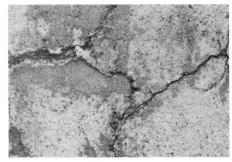

(a) Rusting   (b) Cracking

Figure 4.38   Corrosion of concrete

5) Carbonation

Carbonation is the process during which calcium hydroxide reacts with carbon dioxide and water to form calcium carbonate which is low bonding, or calcium hydrogen carbonate which is soluble in water. Carbonation also causes volume shrinkage of the structure. Carbonation reduces the pH value of concrete and can be easily tested using a phenolphthalein color indicator. Concrete strength decreases with the increase of carbonation depth. The carbonation depth and the rebound value are both considered to estimate the strength of in-situ concrete.

6) Alkali-aggregate reaction

An alkali-aggregate reaction is a swelling reaction between highly alkaline cement paste and reactive silica in aggregates given sufficient moisture, causing aggregate to shape cracking. There are three prerequisites for alkali-aggregate reaction. Firstly, there is excess alkali in cement, specifically, the content ($Na_2O + 0.658K_2O$) is higher than 0.6%. Secondly, the alkali active aggregate content is higher than 1%. Thirdly, the humidity is higher than 80 %. Therefore, measures to reduce the risk of the alkali-aggregate reaction include using inactive aggregates, limiting the content of alkali in cement, and using supplementary materials which consume some alkali. The reaction formula of the alkali-aggregate reaction is shown in Equation (4.7), in which R means Na or K.

$$2R\text{OH} + n\text{SiO}_2 \Longleftrightarrow R_2\text{O} \cdot n\text{SiO}_2 \cdot \text{H}_2\text{O} \tag{4.7}$$

4.4 Admixtures

An admixture is an ingredient of concrete and is added during mixing to improve the properties of fresh, early age, and hardened concrete. There are different types of admixtures for different objectives such as improving workability, strength, and durability, adjusting setting time, etc. Water reducers or plasticizers are to improve concrete workability. Accelerators, retarders, and hydration controllers can be used to control the setting of concrete. Air entrainers, corrosion inhibitors, and alkali-silica reactivity inhibitors can be used to improve durability. Shrinkage reducers are to reduce the shrinkage. Admixtures can be either liquid form or powder form.

4.4.1 Water Reducers

Water reducers can improve the distribution of cement in water mainly through adsorption and dispersion. As shown in Figure 4.39, cement grains develop the static electric charge on their surface as a result of the cement-grinding process and tend to cluster together. The molecules of the water reducer have both positive and negative charges at one end, and a single charge (usually negative) at the other end. These molecules are attracted by the charged surface of the cement grains. The mutual repulsion of like charges pushes the cement grains apart, achieving a better distribution of particles. There are three objectives for using the water reducer:

(1) When the mix proportion is held constant, the flowability is greatly improved.

(2) When the water content is reduced and the workability is held constant, the strength can be greatly improved.

(3) When the content of water and cement is reduced, the strength and workability are held constant, and the amount of cement and the cost of the mix can be greatly reduced.

Civil Engineering Materials

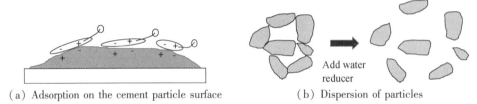

(a) Adsorption on the cement particle surface   (b) Dispersion of particles

Figure 4.39 Effects of the water reducers

4.4.2 Air Entrainer

Air entrainers produce tiny air bubbles in the hardened concrete to provide space for water to expand during freezing. Similar to the water reducer, the hydrophobic tail of the air entrainer is attracted to air, and surface tension is reduced, encouraging the formation of bubbles. The charge around the bubble leads to repulsive forces between bubbles. As shown in Figure 4.40, in air-entrained concrete, air bubbles provide space for water during freezing, relieving hydrostatic pressure in concrete. The entrapped air voids have a diameter of 1 mm or larger and represent approximately 0.2% to 3% of the concrete volume. The entrained air voids are not connected and therefore concrete impermeability is not compromised.

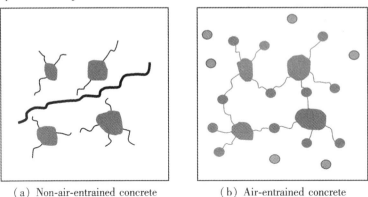

(a) Non-air-entrained concrete   (b) Air-entrained concrete

Figure 4.40 Effects of air entrainers in concrete

4.4.3 Setting Adjuster

Accelerators are used to develop the early strength of concrete at a faster rate. Calcium chloride is the most widely used accelerator. However, it also increases the risks of corrosion and therefore should be cautiously used. The retarder is to slow down the rate of early hydration of C_3S and C_3A by forming a coating on their particle surface (Figure

4.41). It is usually used to offset the effect of hot weather, to allow unusual placement or long haul distance, and to provide time for special finishes. For example, to build the concrete with exposed aggregates for decoration or a skid-resistant surface that is great for sidewalks, we can use retarder to slow down the setting of cement paste so that we can have sufficient time to remove the surface cement paste as finishing.

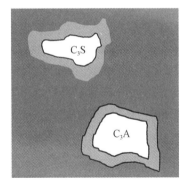

Figure 4.41 Mechanism of setting adjuster

4.4.4 Shrinkage Reducer

Shrinkage reducers can be used to reduce drying shrinkage through expansion compensation. As shown in Figure 4.42, with the addition of the expansion compensating shrinkage reducer, concrete expands a little in the first several days and then the long-term shrinkage is greatly reduced.

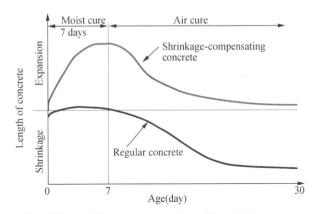

Figure 4.42 Effects of the shrinkage reducer by shrinkage-compensation

There are also other types of shrinkage reducers treating the "cause" of drying shrinkage within the capillary and pore of the cement paste. The shrinkage-reducing admixtures operate by interfering with the surface chemistry of the air/water interface within the capillary or pore, reducing surface tension effects and consequently reducing the shrinkage as water evaporates from within the concrete. Those shrinkage-reducing concrete mixtures are typically comprised of polyoxyalkylene alkyl ether or similar composition. When such an admixture is added to concrete during the batching stage, it can significantly reduce both the early and long-term drying shrinkage by 30% to 50%.

4.5 Mix Design

The objective of cement concrete mix design is to determine the proportion of cement, water, fine and coarse aggregates. The key parameters used in mix design are water content m_{w0}, water-cement ratio (W/C), and sand ratio β_x, based on which the proportion of the four constitutes can be calculated. The mix proportion is usually presented either as the weight per cubic meter or the relative weight ratio. The concrete mix design generally includes 6 steps, i.e. information collection, initial formula, basic formula, lab formula, field formula, and water reducer adjustment. An example is presented as follows to go through the whole procedure.

4.5.1 Information Collection

The first step of mix design is to collect the information on the requirements of designed concrete and the properties of materials:

(1) The designed strength grade of the concrete is C40. It is used in a heavily reinforced section in a level A dry environment. The design life is 50 years. An automatic casting device is used for concrete handling and casting.

(2) The cement is a Grade 42.5 ordinary Portland cement with a density of 3150 kg/m^3 and a safety factor of 1.16.

(3) Natural river sand with a density of 2650 kg/m^3 and a fineness modulus of 2.72 is used as the fine aggregate. The moisture content of the sand is 3%.

(4) A crushed stone with a maximum size of 40 mm and a density of 2700 kg/m^3 is used as the coarse aggregate. The moisture content of the coarse aggregate is 1%.

4.5.2 Initial Formula

The initial formula is determined based on the collected information and the specifications. To obtain the proportion of the four constitutes, this procedure includes nine steps:

(1) Steps 1 and 2 are to determine the slump and water content.

(2) Steps 3 to 6 are to determine concrete strength, cement strength, water-cement ratio, and cement content. Step 7 is to check the content of water and cement for durability.

(3) Steps 8 and 9 are to determine the sand ratio and calculate the content of fine and coarse aggregates.

1) Determining the slump

According to the specification JGJ 55—2011, the slump is determined based on the shape and reinforcement of the structure and the handling method as shown in Table 4.5. This example is a heavily reinforced section with machine handling, the slump should be between 50 and 70 mm.

Table 4.5 Slump requirements

Structures	Handling	
	Equipment (mm)	Manual (mm)
Foundation or ground	0 – 30	20 – 40
No or less reinforced sections	10 – 30	30 – 50
Slabs, beams, and pillars	30 – 50	50 – 70
Heavily reinforced	50 – 70	70 – 90
Very heavily reinforced	70 – 90	90 – 120

2) Determining the water content

The water content is determined based on the slump and the maximum size and type of coarse aggregates. Table 4.6 recommends the water content for medium sand with a fineness modulus between 2.3 and 3.0. The water content should be reduced by 5 to 10 kg/m^3 for fine sand and increased by 5 to 10 kg/m^3 for coarse sand. Table 4.6 is also for a water-cement ratio between 0.4 and 0.8. If the water-cement ratio is lower than 0.4,

Table 4.6 Determination of water content (kg/m^3)

Workability		Maximum size of gravel (mm)				Maximum size of crushed stone (mm)			
Indexes	Value	10	20	31.5	40	16	20	31.5	40
Vebe consistency (s)	16 – 20	175	160	—	145	180	170	—	155
	11 – 15	180	165	—	150	185	175	—	160
	5 – 10	185	170	—	155	190	180	—	165
Slump (mm)	10 – 30	190	170	160	150	200	185	175	165
	35 – 50	200	180	170	160	210	195	185	175
	55 – 70	210	190	180	170	220	205	195	185
	75 – 90	215	195	185	175	230	215	205	195

the water content should be determined based on laboratory trials. For the slump higher than 90mm, 5 kg/m³ more water should be added for each additional 20 mm slump. In this example, the coarse aggregate is a crushed stone with a maximum aggregate size of 40 mm and the slump is between 50 and 70 mm, and therefore the water content m_{w0} is determined as 185 kg/m³.

3) Determining designed concrete strength

The average compressive strength of our designed concrete should be higher than the concrete grade or the characteristic compressive strength. It should be calculated by Equation (4.1) for the conventional concrete with a strength grade lower than 60 MPa. When no historical test results are available to calculate standard deviation, the specification JGJ 55—2011 recommends standard deviations σ as shown in Table 4.7. In this example, the characteristic compressive strength $f_{cu,k}$ of the designed concrete is 40 MPa, and therefore the designed strength f_{cu} is calculated as (rounding to one decimal place):

$$f_{cu} = f_{cu,k} + 1.645\sigma = 40 + 1.645 \times 5.0 \approx 48.2 (\text{MPa})$$

Table 4.7 Recommended standard deviations for different concrete grades

Concrete strength grades	C7.5 – C20	C20 – C45	C50 – C60
Standard deviation σ (MPa)	4.0	5.0	6.0

4) Determining the cement strength

The actual cement strength f_{ce} is to use a safety factor γ_c to time the cement strength grade $f_{ce,k}$ (rounding to one decimal place). When no safety factor is available, the mix design specification JGJ 55—2011 also recommends the safety factors as shown in Table 4.8.

$$f_{ce} = \gamma_c f_{ce,k} = 1.16 \times 42.5 \approx 49.3 (\text{MPa})$$

Table 4.8 Safety factors for different cement grades

Cement strength grades	32.5	42.5	52.5
Safety factors	1.12	1.16	1.10

5) Determining the water-cement ratio

Based on the strength of cement and concrete, and the aggregate type, the water-cement ratio can be calculated by Equation (4.5) and Table 4.5. In this case, the water-cement ratio is calculated as (rounding to two decimal places):

$$\frac{W}{C} = \frac{Af_{ce}}{f_{cu,0} + ABf_{ce}} = \frac{0.53 \times 49.3}{48.2 + 0.53 \times 0.20 \times 49.3} \approx 0.49$$

6) Determining the cement content

The cement content is calculated based on the water-cement ratio and water content (rounding to the nearest integer):

$$m_c = \frac{m_{w0}}{W/C} = \frac{185}{0.49} \approx 378 \,(\text{kg})$$

7) Checking durability requirements

To ensure sufficient durability of the structure for different environment levels and design lives, the specification GB/T 50476—2019 provides the requirements of the minimum concrete strength grade, maximum water-cement ratio, and minimum cement content as shown in Table 4.9. In this example, the design life is 50 years, and the environment level is A. The strength grade, water-cement ratio, and cement content are C40, 0.49, and 378 kg/m^3, respectively, and all meet the requirements of C25, 0.60, 260.

Table 4.9 Durability requirements of the minimum strength, maximum water-cement ratio, and minimum cement content

Environment level	Design life (Year)		
	100	50	30
A	C30,0.55,280	C25,0.60,260	C25,0.65,240
B	C35,0.50,300	C30,0.55,280	C30,0.60,260
C	C40,0.45,320	C35,0.50,300	C35,0.50,300
D	C45,0.40,340	C40,0.45,320	C40,0.45,320
E	C50,0.36,360	C45,0.40,340	C45,0.40,340
F	C55,0.33,380	C50,0.36,360	C50,0.36,360

8) Determining the sand ratio

The sand ratio is mainly determined based on the workability considering the water-cement ratio and aggregate type and size as shown in Table 4.10. In this example, the

Table 4.10 Sand ratio

Water-cement ratio	Maximum size of gravel (mm)			Maximum size of the crushed stone (mm)		
	10	20	40	16	20	40
0.40	26 – 32	25 – 31	24 – 30	30 – 35	29 – 34	27 – 32
0.50	30 – 35	29 – 34	28 – 33	33 – 38	32 – 37	30 – 35
0.60	33 – 38	32 – 37	31 – 36	36 – 41	35 – 40	33 – 38
0.70	36 – 41	35 – 40	3 – 39	39 – 44	38 – 43	36 – 41

maximum size of the crushed stone is 40 mm and the water-cement ratio is 0.49, and therefore the sand ratio β_x is determined as 33%.

9) Determining the proportion of sand and aggregates

The total volume of the four materials plus the content of air voids α equals 1 m³. Since no air entrainer is used, we can assume the content of air voids α as 1%. The amount of sand and coarse aggregates can be calculated by solving Equation (4.8).

$$\begin{cases} \dfrac{m_c}{\rho_c} + \dfrac{m_w}{\rho_w} + \dfrac{m_s}{\rho_s} + \dfrac{m_g}{\rho_g} + \alpha = 1 \\ \dfrac{m_{s0}}{m_{s0} + m_{g0}} = \beta_x \end{cases} \quad (4.8)$$

Based on the obtained value, m_{s0} and m_{g0} can be calculated as (rounding to the nearest integer).

$$\begin{cases} \dfrac{378}{3150} + \dfrac{185}{1000} + \dfrac{m_{s0}}{2650} + \dfrac{m_{g0}}{2700} + 0.01 = 1 \\ \dfrac{m_{s0}}{m_{s0} + m_{g0}} = 0.33 \end{cases}$$

$$\begin{cases} m_{s0} \approx 608 \text{ (kg)} \\ m_{g0} \approx 1232 \text{ (kg)} \end{cases}$$

The initial formula for 1 m³ concrete can be summarized as shown in Table 4.11.

Table 4.11 The proportion of sand and aggregates

Materials	Cement	Water	Sand	Aggregates
Weight (kg/m³)	378	185	608	1232
Proportion	1	0.49	1.61	3.26

4.5.3 Basic Formula

The basic formula is obtained by checking the workability through trial tests and adjusting the content of cement paste. In this example, we mix 20 L concrete and test the workability. The slump is 28mm, less than the designed range of 50 – 70 mm. The formula needs adjustment by fixing the water-cement ratio and adding cement paste. After adding 5% more cement and water, the slump is 65mm, meeting the requirements. The cohesiveness and water retention also meet the requirements. Therefore, the workability of the new adjusted formula is acceptable.

To calculate the adjusted formula, we first calculate the material proportion for the 20 L mix (rounding to two decimal places).

$$\begin{cases} m_c = 378 \times 0.02 = 7.56 \text{ (kg)} \\ m_w = 185 \times 0.02 = 3.7 \text{ (kg)} \\ m_s = 608 \times 0.02 = 12.16 \text{ (kg)} \\ m_g = 1232 \times 0.02 = 24.64 \text{ (kg)} \end{cases}$$

After adding 5% more cement and water, the mix proportion for the 20 L mix is

$$\begin{cases} m'_c = 7.56 \times (1 + 0.05) \approx 7.94 \text{ (kg)} \\ m'_w = 3.7 \times (1 + 0.05) \approx 3.89 \text{ (kg)} \\ m_s = 608 \times 0.02 = 12.16 \text{ (kg)} \\ m_g = 1232 \times 0.02 = 24.64 \text{ (kg)} \end{cases}$$

The ratio of the mix now is (rounding to two decimal places)

cement: water: sand: coarse aggregates = 1: 0.49: 1.53: 3.10

The cement content can be calculated by solving the following Equation

$$\frac{m_{cl}}{3150} + \frac{0.49 m_{cl}}{1000} + \frac{1.53 m_{cl}}{2650} + \frac{3.1 m_{cl}}{2700} + 0.01 = 1$$

The basic formula is (rounding to the nearest integers)

$$\begin{cases} m_{cl} \approx 393 \text{ (kg/m}^3\text{)} \\ m_{wl} \approx 193 \text{ (kg/m}^3\text{)} \\ m_{sl} \approx 601 \text{ (kg/m}^3\text{)} \\ m_{gl} \approx 1218 \text{ (kg/m}^3\text{)} \end{cases}$$

4.5.4 Lab Formula

The lab formula is obtained by checking the strength through trial tests and adjusting the water-cement ratio. This step involves preparing specimens based on the basic formula and testing the compressive strength. It is usual to fix water content and prepare 3 groups of concrete with the water-cement ratio and water cement ratio ±0.05 or 0.1. The relationship between 28-day concrete strength and water-

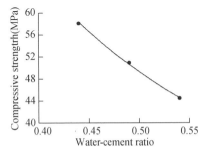

Figure 4.43 Relationship between 28-day strength and water-cement ratio

cement ratio can be obtained as shown in Figure 4.43, based on which the adjusted water-cement ratio can be determined. It can be seen that the water-cement ratio for the 48.2 compressive strength is 0.51. The water content is 193 g/m³, and the mix proportion can be determined by solving the following equation (rounding to the nearest integer).

$$\begin{cases} \dfrac{193/0.51}{3150} + \dfrac{193}{1000} + \dfrac{m'_{s2}}{2650} + \dfrac{m'_{g2}}{2700} + 0.01 = 1 \\ \dfrac{m'_{s2}}{m'_{s2} + m'_{g2}} = 0.33 \end{cases}$$

$$\begin{cases} m'_{c2} \approx 379 \,(\text{kg/m}^3) \\ m'_{w2} \approx 193 \,(\text{kg/m}^3) \\ m'_{s2} \approx 600 \,(\text{kg/m}^3) \\ m'_{g2} \approx 1218 \,(\text{kg/m}^3) \end{cases}$$

Before submitting the lab formula to the job site or concrete plant, we need to adjust the formula with the apparent density because the sum of the adjusted weight of the four materials might not be exactly 1 m³. The apparent density of fresh concrete can be determined by filling the concrete into a rigid container with known volume and measuring the weight. The apparent density in this example is 2427 kg/m³. The final lab formula expressed as the weight per 1 m³ is calculated as (rounding to the nearest integer):

$$\begin{cases} m_{c2} = 2427 \times \dfrac{379}{379 + 193 + 600 + 1218} \approx 385 \,(\text{kg/m}^3) \\ m_{w2} = 2427 \times \dfrac{193}{379 + 193 + 600 + 1218} \approx 196 \,(\text{kg/m}^3) \\ m_{s2} = 2427 \times \dfrac{600}{379 + 193 + 600 + 1218} \approx 609 \,(\text{kg/m}^3) \\ m_{g2} = 2427 \times \dfrac{1218}{379 + 193 + 600 + 1218} \approx 1237 \,(\text{kg/m}^3) \end{cases}$$

4.5.5 Field Formula

The field formula is to adjust the lab formula by considering the moisture content of sand and coarse aggregates on the job site. Less water should be used considering the moisture and more sand and coarse aggregates on the job site should be used to compensate for the weight of moisture. The field formula can be calculated and used in the plant for concrete mix production as follows:

$$\begin{cases} m_{c3} = 385\,(\text{kg}/\text{m}^3) \\ m_{w3} = 196 - 609 \times 0.03 - 1237 \times 0.01 \approx 165\,(\text{kg}/\text{m}^3) \\ m_{s3} = 609 \times (1 + 0.03) \approx 627\,(\text{kg}/\text{m}^3) \\ m_{g3} = 1237 \times (1 + 0.01) \approx 1249\,(\text{kg}/\text{m}^3) \end{cases}$$

4.5.6 Water Reducer Adjustment

When a water reducer is used, the lab formula needs to be adjusted. If we add a water reducer with an 18% reducing rate in this formula and maintain the slump and strength the same. The content of water and the content cement are both reduced. The adjusted water content is calculated based on the water reduction rate and the cement content is calculated based on the water-cement ratio (rounding to the nearest integer).

$$m_{w2\_ad} = 196 \times (1 - 0.18) \approx 161\,(\text{kg}/\text{m}^3)$$
$$m_{c2\_ad} = 161/0.51 \approx 316\,(\text{kg}/\text{m}^3)$$

The apparent density of the concrete with a water reducer is 2400 kg/m³. The adjusted lab formula expressed as the weight per 1 m³ is calculated as:

$$\begin{cases} m_{w2a} = 196 \times (1 - 0.18) \approx 161\,(\text{kg}/\text{m}^3) \\ m_{c2a} = \dfrac{161}{0.51} \approx 316\,(\text{kg}/\text{m}^3) \\ m_{s2a} = 2400 \times \dfrac{609}{161 + 316 + 609 + 1236} \approx 629\,(\text{kg}/\text{m}^3) \\ m_{g2a} = 2390 \times \dfrac{1236}{161 + 316 + 609 + 1236} \approx 1272\,(\text{kg}/\text{m}^3) \end{cases}$$

Questions

1. Define the three workability characteristics of fresh concrete.
2. Why is the sand ratio critical for the workability of fresh concrete?
3. Define the ITZ of cement concrete.
4. Why the axial compressive strength is usually lower than the cubic compressive strength?
5. Discuss the change in the volume of concrete at early ages.
6. Discuss creep and relaxation of concrete.
7. The four-point loading flexure strength test was performed on a concrete beam

having a cross-section of 0.15 m by 0.15 m and a span of 0.45 m. If the load at failure was 31.7 kN, calculate the flexure strength of the concrete.

8. Discuss plastic shrinkage and carbonation shrinkage of concrete.

9. Briefly explain the mechanism of the water reducer, air entrainer, setting retarder, and shrinkage reducer.

10. Design the concrete mix according to the following conditions:

(1) The designed strength grade of the concrete is C30. It is used in a heavily reinforced section in a level A dry environment. The design life is 100 years. An automatic casting device is used for concrete handling and casting.

(2) The cement is a Grade 32.5 ordinary Portland cement with a density of 3050 kg/m^3 and a safety factor of 1.12.

(3) Natural river sand with a density of 2510 kg/m^3 and a fineness modulus of 2.65 is used as the fine aggregates. The moisture content of the sand is 4%.

(4) A crushed stone with a maximum size of 40 mm and a density of 2650 kg/m^3 is used as the coarse aggregates. The moisture content of the coarse aggregates is 2%.

5 Inorganic Binder Stabilized Materials

The inorganic binder stabilized material (IBSM) is a general term that applies to an intimate mix of soils and/or aggregates with an inorganic binder such as cement, lime, fly ash, or other cementitious materials, and water that hardens after compaction and curing to form a strong and durable paving material. The crushed or undisturbed soils can be classified into fine, medium, and coarse-grained soils. Different types of soils and inorganic binders are mixed to obtain various IBSMs, such as the lime stabilized soil, lime and fly ash stabilized soil/aggregate/macadam, cement stabilized soil/aggregate/macadam, cement-treated aggregate.

Due to the good stability and high stiffness, the IBSM is widely used as the base or subbase of asphalt and cement concrete pavements. The pavement base layer built with the IBSM is called an inorganic binder stabilized base (IBSB). It is also called a semi-rigid base because its stiffness is significantly higher than that of the flexural pavement base such as asphalt-treated base or aggregate base, but lower than the typical rigid pavement such as cement concrete. More than 95% of asphalt and cement concrete pavements in China utilize the IBSM as the base or subbase and extensive studies have been conducted on their performance. This chapter introduces the modification mechanism, strength, fatigue, shrinkage, and mix design of IBSM.

5.1 Applications and Classifications

5.1.1 Applications

In asphalt pavements, IBSB provides a stiffer and stronger base than an unbound aggregate base or asphalt base and reduces the deflection due to traffic loads, resulting in lower tensile stress at the bottom of the asphalt layer. It improves the ability of the asphalt surface layer to resist fatigue damage. The asphalt surface layer on an IBSB may not encounter fatigue damage with a strong interlayer bonding and the fatigue resistance of the asphalt pavement with IBSB is then determined by the fatigue life of IBSB. Rutting potential is also reduced in an asphalt pavement with an IBSB due to its high stiffness and load distribution capability. The strong uniform support provided by IBSB also reduces stress on top of the subgrade. A thinner IBSB can reduce subgrade stress more than a thicker layer of flexural or unbound aggregate base (Figure 5.1). Subgrade failure and pavement roughness are thus reduced. The slab-like characteristics of IBSB are unmatched by the unbound aggregate base that can fail when the interlock is lost. In addition, IBSB keeps water out and maintains higher levels of strength, even when saturated, thus reducing the potential for pumping of subgrades.

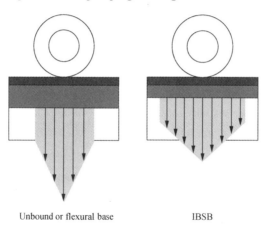

Unbound or flexural base   IBSB

Figure 5.1 Comparison between unbound or flexural base and IBSB

However, a concern for asphalt pavements with IBSB is the reflective cracking which is typically caused by the propagation of the shrinkage cracking of IBSB into the asphalt surface layer. Wide reflective cracks cause severe damage to pavements, such as pumping as the infiltration of moisture. The stiffer IBSB also provides strong support for the concrete pavement. However, moisture may infiltrate into the IBSB through the joints of concrete pavements, increasing the risk of pumping, scouring, and faulting of slabs under repeated traffic loads or freeze-thaw cycles. Therefore, in humid and rainy areas, a drainage base layer with porous open-graded cement stabilized macadam, asphalt stabilized macadam, or crushed stone can be used.

5.1.2 Classifications

1) Cement stabilized material (CSM)

The CSM uses Portland cement to treat and stabilize soils and/or aggregates as pavement bases or subbases to meet specific strength requirements. The aggregates or soils can be (1) any combination of gravel, stone, sand, silt, and clay; (2) miscellaneous materials such as caliche, scoria, slag, sand-shell, cinders, and ash; (3) waste materials from aggregate production plants; (4) high-quality crushed stone and gravel base course aggregates; or (5) recycled pavements including the pulverized bituminous surface and aggregate base course. According to the soil particle size, CSM can be divided into:

(1) Cement stabilized fine-grained soil, in which the maximum particle size is less than 9.5 mm, and the content of <2.36 mm particles should be no less than 90%.

(2) Cement stabilized medium-grained soil, in which the maximum particle size is less than 26.5 mm, and the content of <19 mm particles should be no less than 90%.

(3) Cement stabilized coarse-grained soil, in which the maximum particle size is less than 37.5 mm, and the content of <31.5 mm particles should be no less than 90%.

Cement stabilized fine-grained soil is also called cement stabilized soil or cement soil. Cement stabilized medium or coarse-grained soil is called cement stabilized sand/gravel/aggregate/macadam depending on the type of soils or aggregates. Lime can be added to produce cement and lime stabilized materials (CLSM). Cement is the main cementing material and aggregates or soils are bonded by the hydration of cement. Similar to cement concrete, CSM continues to gain strength with the increase in curing time. The compressive strength of CSM cores can be as high as 30 MPa after 15 years of service (Figure 5.2). CSM provides a relatively strong, stiff, and durable slab structure, compared with other IBSM. However, the high dry and thermal shrinkage and cracking are the drawbacks of CSM, which are more severe when the content of fine particles and cement is high.

Civil Engineering Materials

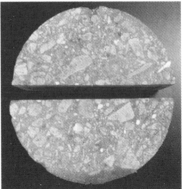

Figure 5.2 Cores of cement-stabilized base material

2) Lime stabilized material (LSM)

The LSM uses lime to treat and stabilize soils and/or aggregates as pavement bases or subbases to meet specific strength requirements. There are different types of LSMs. Lime stabilized fine-grained soil is called lime stabilized soil or lime soil. Lime stabilized medium or coarse-grained soil is called lime stabilized sand/gravel/aggregate/macadam depending on the type of soils or aggregates. Lime is often used with other binders such as cement and industrial wastes such as fly ash, slag, coal cinders. When used with cement, the lime acts as an activator. A small number of chemical additives such as $CaCl_2$, NaOH, Na_2CO_3. can be added to improve the interaction between lime and soils.

The LSM has good mechanical properties, fair water stability, and freeze-thaw resistance. Its initial strength is low but the strength will increase with the increase in curing age or service time. The compressive strength of LSM cores can be as high as 15 MPa after 20 years of service. It also has the drawback of large shrinkage and is prone to crack. LSM is suitable for the base of all classes of pavements and the subbase of roads of Class II or lower classes, but should not be used as the base of the expressway. In the moist sections of pavements in permafrost areas or excessively humid areas, LSM is not recommended and measures should be taken to prevent water from infiltrating the LSM base.

3) Lime and industrial waste stabilized material (LIWSM)

The industrial wastes include fly ash, blast furnace slag, aged steel slag, coal gangues, and other cementitious materials that can be used with lime together to treat and stabilize soils and/or aggregates to meet specific strength requirements. LIWSM can be divided into lime and fly ash stabilized materials (LFASM) and lime and other wastes stabilized materials. Lime and fly ash stabilized fine/medium/coarse-grained soil is the most widely used. The industrial wastes can also be used with cement to produce the cement and industrial waste stabilized materials (CIWSM) and the cement and fly ash stabilized materials (CFASM), in which the industrial wastes can have pozzolanic reactions with the calcium hydroxide produced during cement hydration to further improve the strength of the mix.

In the LIWSM, lime is the activator and industrial wastes are the main cementing materials. The early-age strength of the LIWSM is low but will increase with time. Therefore, the LIWSM base can provide a strong and durable slab structure. With the change of temperature and moisture conditions, the LIWSM is also prone to crack, especially for those with a high content of fine particles and binders. The resistance to moisture damage of LIWSM is worse than that of the same material stabilized with cement, but its drying and thermal shrinkage coefficients are smaller.

5.2 Modification Mechanism

5.2.1 Cement Stabilized Material

The stabilization process of CSM includes complex reactions and interactions between cement, soils, and water, resulting in significant variations of mixed properties. Chemical reactions include cement hydration, polymerization of soil organic matter, reactions between cement hydration products and clay minerals, etc. Physicochemical interactions include the adsorption between clay particles and cement or cement hydration products, agglomeration effect, diffusion and osmosis of water and hydration products, dissolution and crystallization of hydration products, etc. Physical interactions include mechanical crushing of soils, mixing, compaction of mixes, etc.

1) Hydration of cement

The cementing capability of cement hydration products is the main source of the strength of CSM. Hydration products form the interlock in the pores of the soil and bond the soil particles (Figure 5.3). The hydration conditions of cement are much worse in CSM than that in concrete due to the lower cement content, less free water, high specific surface area and hydrophilicity of soil particles, strong adsorption of soil to cement hydration products, and the acid environment in some soils. Because clay minerals have strong adsorption with $Ca(OH)_2$ in hydration products, the alkalinity of the hydration solution is reduced, influencing the stability of hydration products. In addition, the ratio of CaO over SiO_2 in the hydrated calcium silicate gradually decreases, and $Ca(OH)_2$ precipitates, which changes the structure and properties of hydration products and influences the mechanical properties of the mix. Therefore, adding lime can improve the alkalinity of the mix and the strength of CSM.

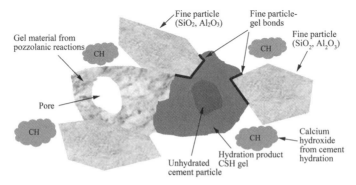

Figure 5.3 Hydration of cement in CSM

With the continuous hydration of cement, the $Ca(OH)_2$ produced during the hydration of cement can react with active silicon and alumina oxides in the soil to produce CSH and calcium aluminate hydrate. This pozzolanic reaction also helps improve the long-term strength of CSM (Figure 5.3). The chemical reactions include:

$$SiO_2 + xCa(OH)_2 + mH_2O \Longleftrightarrow xCaO \cdot SiO_2 \cdot (x+m)H_2O$$
$$Al_2O_3 + xCa(OH)_2 + mH_2O \Longleftrightarrow xCaO \cdot Al_2O_3 \cdot (x+m)H_2O \quad (5.1)$$

2) Cation exchange

Clay particles in soil have high activity due to their small size and large specific surface area. When clay particles encounter water, their surface is usually negatively charged, forming an electric field. This layer of negatively charged ions is called potential ions. The surface of clay particles with a negative charge attracts and absorbs cations such as K^+, Na^+. from the surrounding solution, forming an electric double layer. The ions with opposite charges of the potential ions are called counter ions. In the electric double layer, potential ions form the inner layer and counter ions form the outer layer. The counter ions close to particles have a tight bond to the particle surface. When clay particles move, tightly bonded counter ions will move with particles, while other counter ions will not move. Therefore, a slip surface generates between the moving and non-moving counter ions.

Cation exchange is the interchange between a cation on the surface of the negatively charged particle (i.e. clay mineral or organic colloid) and the soil solution. Portland cement provides sufficient Ca^{2+} in the solution during hydration. The Ca^{2+} has higher electrovalency than K^+, Na^+, etc., and higher attractions with potential ions. The Ca^{2+} also has a rapid potential reduction rate of the electric double layer. Therefore, Ca^{2+} reduces the electric potential and the thickness of the electric double layer, causing clay particles to get closer and agglomerate. It will change the plasticity of the soil and make the soil have a specific strength and stability (Figure 5.4).

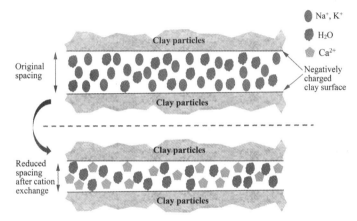

Figure 5.4 Cation exchange

The cation exchange causes the restructuring of modified soil/aggregate particles and changes the texture of the material from that of a plastic, fine-grained material to a more friable, granular soil/aggregate. This process includes flocculation and agglomeration. Flocculation, which is the process of soil particles altering

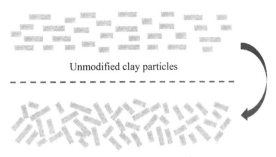

Figure 5.5 Chemical excitation

their arrangement from a flat parallel structure to a more random edge-to-face orientation (Figure 5.5). Agglomeration refers to the weak bonding at the edge-surface interfaces of the soil particles, which forms larger aggregates from finely divided soil particles and further improves the texture of the soil/aggregate.

3) Chemical excitation

Calcium ions not only influence the structure of an electric double layer of clay particles, but also change the chemical properties of clay in an alkaline environment. The main mineral in the soil is aluminosilicate, which contains large quantities of silicon-oxygen tetrahedrons and alumina-oxygen octahedrons. In general, these minerals are relatively stable until the pH value of the soil solution increases to a specific level. The $Ca(OH)_2$ generated during cement hydration dissolves and increases the alkalinity of the soil solution. Then, the activity of SiO_2 and Al_2O_3 in the clay minerals will be excited to react with Ca^{2+} to form calcium silicates and calcium aluminates, such as $4CaO \cdot 5SiO_2 \cdot 5H_2O$, $4CaO \cdot Al_2O_3 \cdot 19H_2O$, $3CaO \cdot Al_2O_3 \cdot 16H_2O$, $CaO \cdot Al_2O_3 \cdot 10H_2O$. These minerals have the similar composition and structure as the hydration products of cement and play a cementing role in the material, the cementing material coats the surface of clay particles and bonds the clay particles together with the cement hydration products. As a result, the excitation of calcium hydroxide on clay minerals will further improve the strength and stability of cement-stabilized soils.

4) Carbonation

$Ca(OH)_2$ generated by cement hydration not only reacts with clay minerals, but also carbonizes with CO_2 in the air with the presence of moisture and forms calcium carbonate crystals. The volumetric expansion generated during the formation of calcium carbonate can also fill and strengthen the soil matrix, although this effect is relatively weak and the reaction process is slow.

$$Ca(OH)_2 + CO_2 + nH_2O \Longleftrightarrow CaCO_3 + (n+1)H_2O \tag{5.2}$$

5.2.2 Lime Stabilized Material

Properties of soils are improved due to the interaction between lime and soils after adding lime. In the initial stage, lime mainly causes agglomeration of soils, reduction of plasticity, an increase in optimum moisture content, and reduction of maximum dry density. In the later stages, the formation of the crystalline structure improves strength and stability. Soils contain clay colloidal particles and have complicated chemical compositions and minerals. Therefore, in addition to physical adsorption, complex physicochemical and chemical interactions occur when lime is added to soils. The degree of those interactions is related to external factors such as humidity, temperature, and varies depending on humidity and temperature. The interactions between lime and soils can be summarized into the following four mechanisms.

1) Cation exchange

Cation exchange in LSM is similar to that in CSM. When lime is added into the soil at the optimum water content, calcium hydroxide or magnesium hydroxide dissolves in water and dissociates into calcium ions, magnesium ions, and hydroxide ions. When the diffusion layer of clay colloidal particles mostly contains K^+ and Na^+, the Ca^{2+} and Mg^{2+} cations are exchanged with the low electrovalency K^+ and Na^+. Both the thickness of the colloidal diffusion layer and the electric potential decrease, the van der Waals force increases, and the particles are bonded more closely. Thus, the agglomerate structure of the lime stabilized soil is improved, reducing wet collapsibility and swelling of the soil. Cation exchange occurs rapidly in the early stage and gradually slows down. It is the main contribution to the early performance of the LSM.

2) Pozzolanic reaction

Similar to the pozzolanic reaction in CSM, the calcium hydroxide in lime reacts with active silicon and alumina oxides in the soil to produce CSH and CAH in LSM. The chemical composition and crystallization structure of generated CSH and CAH are related to the ratio of CaO over SiO_2 or CaO over Ai_2O_3, temperature, and humidity. The bond between soil particles is enhanced due to the formation of the products from the pozzolanic reaction, as well as the symbiosis of crystals. The symbiosis of crystals plays a cementing role in the soil, thus enhancing the strength and water stability of LSM and contributing to strength growth over a considerable period.

3) Carbonation

When lime is added to the soil, calcium hydroxide absorbs water and carbon dioxide from the air to form insoluble calcium carbonate, which is a crystal with high strength and good water stability. The symbiosis of calcium carbonate crystals and soil particles plays a cementing role in the soil. In addition, the volume of the solid phase of calcium carbonate is slightly higher than that of calcium hydroxide during the carbonation reaction, which makes the mix more compacted and stronger. Carbonation reaction is a slow process and depends on the content of CO_2. CO_2 may infiltrate into the mix through pores, rainwater, or be generated by the soil in small amounts. When a layer of calcium carbonate is formed on the surface of particles, it prevents further penetration of CO_2, slowing down the carbonation process. The carbonation of calcium hydroxide lasts for a long time, contributing to the later strength of the lime stabilized soil.

4) Crystallization of calcium hydroxide

With the evaporation of water and the lime-soil reaction, $Ca(OH)_2$ crystals precipitate from the supersaturated solution, resulting in the crystallization of calcium hydroxide. The reaction is shown in Equation (5.3). The crystals bond with soil particles to form strong structures. The solubility of the crystallized $Ca(OH)_2$ is small (almost half of that of amorphous $Ca(OH)_2$), and therefore the strength and water stability of the lime-stabilized soil are improved.

$$Ca(OH)_2 + nH_2O \Longleftrightarrow Ca(OH)_2 \cdot nH_2O \qquad (5.3)$$

In summary, the major strength improvements of LSM are cation exchange and chemical reaction between lime and soils. The cation exchange, which occurs in a short period after lime is added, is the main reaction in the early stage. Then, the pozzolanic reaction, carbonation, and crystallization of calcium hydroxide generate calcium carbonate, CSH, CAH, and calcium hydroxide crystals with different crystallization degrees and structures. As the formation of a large number of crystals, the crystals cross and conjoin each other, forming a strong network of crystalline structures between the soil particles, and the structure of LSM is further improved. The crystalline structure differs from the cementing structure because interaction forces between particles are not intermolecular but chemical bonding forces, which makes them much stronger. The crystalline network does not have thixotropic recovery properties after disruption due to variation of interactions. The chemical reaction of lime has more significant effects on strength. Pozzolanic reaction continues to obtain strength over a long period, and its effect varies greatly with the types of soils and the climate. The slowest reaction is the carbonation of calcium hydroxide, which is one of the main reasons for the strength of the lime-soil to continue to increase in the later period.

5.2.3 Lime and Industrial Waste Stabilized Material

The stabilization mechanism of LIWSM is similar to LSM but varies depending on the added industrial wastes. The main industrial waste material used in LISWM is fly ash containing a large amount of active SiO_2, Al_2O_3, and CaO. It is usually used with lime to stabilize soils through hydration, pozzolanic reaction, cation exchange, carbonation, etc. More details of the production, properties, and composition of fly ash are introduced below.

5　Inorganic Binder Stabilized Materials

1) Production of fly ash

Fly ash is a byproduct discharged from coal-fired power plants. A specific degree of fineness of pulverized coal burns at a high temperature of 1100 – 1600 ℃ and its non-combustible part is discharged with the tail gas. The fine ash collected by the dust collector is called fly ash. In high-temperature combustion, the clayey minerals contained in the pulverized coal are melted and form liquid droplets. When the pulverized coal is discharged from the furnace with the tail gas, it is cooled rapidly to form spherical glass particles with the size of 1 – 50 μm.

The molten glass particles form hollow glass balls because some substances in the pulverized coal decompose and volatilize to produce gas. Sometimes the walls of the glass spheres are honeycombed. The density of fly ash is much less than that of other minerals with the same composition, usually 1900 – 2400 kg/m^3. The bulk density is 600 – 1000 kg/m^3, and the specific surface area is 2700 – 3500 cm^2/g. High calcium fly ash has a relatively higher density of 2500 – 2800 kg/m^3, and a higher bulk density of 800 – 1200 kg/m^3. In addition to the hollow glass ball, fly ash also contains large particles with open pores, small particles wrapped in the hollow glass ball, and grape-shaped particles which are formed from the collision of the small and dense particles in the flue gas.

The colors of fly ash vary from gray to black due to different carbon content and iron content, and its appearance and color are similar to those of cement. According to different dust collection methods, the production of fly ash is classified into the dry process and the wet process. The dry process uses compressed air to transport the fly ash collected by the dust collector, while the wet process uses water to transport. Dry-processed fly ash has relatively fine particles and high activity, while wet-processed fly ash has relatively coarse particles and low activity because part of the active components has hydrated first.

2) Particle properties

The quality of fly ash is greatly influenced by the properties of particles. Particle size distribution of fly ash relates to the type of raw coal, fineness of pulverized coal, and combustion conditions. When fly ash is added to cement paste, the spherical particles in fly ash can play a lubricating role in the cement paste, reducing water consumption. The higher content of spherical particles with a smooth surface in fly ash, the less water content is required. In addition, when the particles of fly ash are finer, their specific surface area is larger, the hydration reaction interface increases, and the activity of fly ash is higher. When the average particle size of fly ash is relatively large with more combined particles, the required water increases, and the activity decreases. It is generally believed that fly ash particles with particle size 5 – 30 μm have higher activity. Most specifications in China require that the residue of fly ash on a 0.08 mm square hole sieve should be less than 8%. The coarse and porous combined particles can be broken up, so that the particle size is reduced and the specific surface area is increased, while finer spherical particles remain in their original shape because they are difficult to grind. The quality of fly ash can be effectively improved by grinding, and particles larger than 0.2 mm in fly ash can be regarded as aggregates.

3) Chemical composition

Fly ash mainly includes silicon and alumina oxides, the total content of which can be higher than 60%. Other chemical composition is shown in Table 5.1. The chemical composition of fly ash is a type of CaO-Al_2O_3-SiO_2 system, and its activity depends on the content of Al_2O_3 and SiO_2. Most specifications in China require that the total content of SiO_2, Al_2O_3, and Fe_2O_3 in fly ash should be more than 70%. CaO only accounts for a small portion of fly ash, but is extremely favorable to the activity of fly ash. Some fly ash can contain up to 34%–45% CaO due to the special raw material and it can hydrate after adding water. In general, fly ash with CaO content below 5% is called low-calcium fly ash; fly ash with CaO content between 5% and 15% is called medium-calcium fly ash; fly ash with CaO content above 15% is called high-calcium fly ash. Calcium can be added during the production of fly ash. ASTM C618-19 defines two classes of fly ash. Class C fly ash contains more CaO and is cementitious, while Class F fly ash contains less CaO and is mainly pozzolanic.

Table 5.1 Chemical composition of fly ash

Composition	SiO_2	Al_2O_3	Fe_2O_3	CaO	MgO	SO_3	Loss on ignition
Content (%)	40 – 60	15 – 40	3 – 10	2 – 5	0.5 – 2.5	<2	1 – 20

4) Hydration of lime and fly ash

The activity of fly ash mainly refers to the pozzolanic property. The natural or artificial pozzolanic materials can not hydrate by themselves, but can react with calcium hydroxide to produce cementitious hydration products in the presence of water and lime. Fly ash and Portland cement are both CaO-Al_2O_3-SiO_2 systems and have many similar properties.

Glass phase in fly ash is also a metastable phase that is formed at high temperatures and therefore tends to transform to a stable phase by hydration under proper conditions.

Fly ash usually cannot self-hydrate due to the low content of CaO. However, fly ash is active when it is excited by quicklime, hydrated lime, cement hydration, gypsum, and other alkaline substances. Particularly, $Ca(OH)_2$ shows an obvious stimulating effect on the hydration activity of fly ash. The primary chemical reactions are as follows.

$$\begin{aligned} &CaO + H_2O \longrightarrow Ca(OH)_2 \\ &Ca(OH)_2 + CO_2 \longrightarrow CaCO_3 + H_2O \\ &Ca(OH)_2 + SiO_2 + H_2O \longrightarrow xCaO \cdot ySiO_2 \cdot zH_2O \\ &Ca(OH)_2 + Al_2O_3 + H_2O \longrightarrow xCaO \cdot yAl_2O_3 \cdot zH_2O \\ &Ca(OH)_2 + SiO_2 + Al_2O_3 + H_2O \longrightarrow xCaO \cdot yAl_2O_3 \cdot zSiO_2 \cdot wH_2O \\ &Ca(OH)_2 + SO_4^{2-} + Al_2O_3 + H_2O \longrightarrow xCaO \cdot yAl_2O_3 \cdot zCaSO_4 \cdot wH_2O \end{aligned} \quad (5.4)$$

5) Requirements for fly ash

The specification JTG/T F20—2015 provides the requirements for fly ash, including the content of carbon, oxides, fineness, moisture, and sulfur.

(1) Carbon content

The carbon content of fly ash is expressed by loss on ignition at 800 – 900 ℃. Large loss on ignition indicates insufficient combustion, which has a great impact on the quality of fly ash. The particles of coal residues are coarse and porous and have high absorption and poor stability. When the carbon content in fly ash is high, the water demand increases, reducing strength and durability, and increasing shrinkage. In addition, unburned pulverized coal will form a hydrophobic layer on the surface to prevent the infiltration of water into fly ash when it meets water, thus influencing the hydration reaction of fly ash. Moreover, powdered carbon is also the cause of volume variation and poor stability of the material. Some states in the US require that loss on ignition of fly ash should not exceed 10%. Fly ash in some places in China has a loss on ignition of up to 18%, but it can still achieve the required strength when stabilizing aggregates and soils together with lime. The specification JTG/T F20—2015 requires that loss on ignition of fly ash should not exceed 20%.

(2) Oxides content

Oxides content refers to the total content of SiO_2, Al_2O_3, and Fe_2O_3 in the fly ash. The content of oxides in fly ash significantly influences the strength of lime-fly ash mixes, and it is required that the content of oxide should be greater than 70 %.

(3) Fineness

The fineness of fly ash particles directly influences the reaction rate and the number of reaction products after mixing with lime and cement. The finer particles of fly ash have a larger specific surface area and higher activity, thus enhancing the strength of the mix. The specification specifies that the specific surface area of fly ash should be greater than 2500 cm^2/g. In addition, the passing percentage at 0.3 and 0.075 mm sieves should be greater than 90 % and 70 %, respectively.

(4) Moisture content

In the process of storage and transportation, fly ash is often sprinkled with water to prevent dust pollution. Water decreases the activity of fly ash and causes agglomeration. Therefore, the moisture content of fly ash should not exceed 35 %.

(5) Sulfur content

The limestone-gypsum wet flue gas desulfurization is adopted to decrease sulfur dioxide in the water absorption tower, causing a significant increase of sulfur content in fly ash (generally greater than 0.3 %). Therefore, the specification GB/T 1596—2017 requires that the content of sulfur trioxide (SO_3) should be less than 3 %, according to the test method of coal ash in specification GB/T 1574—2007.

5.3 Mechanical Properties

5.3.1 Unconfined Compression Test

As specified in JTG E51—2009 T0805, the objective of the unconfined compression test is to determine the required binder content of IBSM.

1) Sample preparation

The test uses a cylinder with a height/diameter ratio of 1:1. It is prepared by the static compaction method. The height or diameter of the specimens for fine, medium, and coarse-grained soils is 50 mm, 100 mm, and 150 mm, respectively. The number of specimens prepared in each group is related to the type of soils and operations. At least 6, 9, and 13 specimens should be prepared per group for fine, medium and coarse-grained soils, respectively.

After determining the optimum water content using the compaction method, water is added into the air-dried soils and then the soils are soaked in an airtight container. For LSM or CLSM, lime can be added and soaked together with the soils. After soaking, a specific amount of cement or lime is added and mixed evenly to prepare the samples. For CSM, sample preparation should be completed within 1 hour. According to the size of the specimen, a specific weight of mixes is weighted, which is $m_1 = \rho_d V(1 + \omega)$, where ρ_d is the dry density of the mix, V is the volume of the specimen, ω is the moisture content. The compression force should be maintained for 1 min. For the soils with cohesiveness, the sample can be extruded after the compaction, while for the soils with less cohesiveness, it is suggested to extrude the sample after a few hours to ensure that the specimen will not collapse during extruding. Then, the specimens should be immediately placed in a sealed moist cabinet or thermostatic chamber to cure at 20 ℃ ±2 ℃.

2) Compressive strength test

The specimen should be immersed in water on the day before the 7-day compressive strength test. The weight of the specimen should be weighed again before immersion. During the curing period, the weight loss of small, medium, and large specimens should be no more than 1 g, 4 g, and 10 g, respectively. After draining visible free water out of the surface of the specimen, and measuring the height and weight of the specimen, a compressive load is applied on the top of the specimen at a 1 mm/min loading rate till the fracture of the specimen. The maximum compressive force P is recorded. A sample is taken from the inner part of the fractured specimen to determine its moisture content. The compressive strength of the specimen is calculated as below

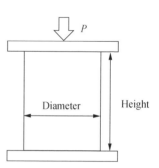

Figure 5.6 Compressive strength test

$$R_c = \frac{P}{A} \tag{5.5}$$

where R_c = compressive strength (MPa);
  P = load at failure (N);
  A = section area of specimen (mm²);

$$A = \frac{1}{4}\pi D^2 \tag{5.6}$$

where D = diameter of specimen (mm).

The average compressive strength $\bar{R}$ and the coefficient of variation C_v can also be obtained. The C_v for small, medium, and large size specimens should be no greater than 10%, 15%, and 20%, respectively. If the test results do not meet the requirements, the test should be repeated and the number of parallel tests should be increased accordingly. The content of inorganic binder corresponding to the minimum average strength $\bar{R}$ satisfying Equation (5.7) is selected as the designed content. The 7-day compressive strength is used as the major design criteria for the mix design of IBSM. Table 5.2 list the strength requirements from specification JTG/T F20—2015.

$$\bar{R} \geqslant \frac{R_d}{1 - Z_\alpha \cdot C_v} \tag{5.7}$$

where R_d = design compressive strength (MPa);
C_v = coefficient of variation of test results;
Z_α = the Z score for different confidence levels α in a standard normal distribution. For expressway and Class I road, $\alpha = 95\%$, and $Z_\alpha = 1.645$. For other classes of roads, $\alpha = 90\%$, and $Z_\alpha = 1.282$.

Table 5.2 Required 7-day unconfined compressive strength for IBSM (MPa)

Materials	Courses	Road classes	Super and very high traffic volume	High traffic volume	Medium and low traffic volume
Cement stabilization	Base	Expressway and Class I road	5.0 – 7.0	4.0 – 6.0	3.0 – 5.0
		Class II and lower class roads	4.0 – 6.0	3.0 – 5.0	2.0 – 4.0
	Subbase	Expressway and Class I road	3.0 – 5.0	2.5 – 4.5	2.0 – 4.0
		Class II and lower class roads	2.5 – 4.5	2.0 – 4.0	1.0 – 3.0
Cement and fly ash stabilization	Base	Expressway and Class I road	4.0 – 5.0	3.5 – 4.5	3.0 – 4.0
		Class II and lower class roads	3.5 – 4.5	3.0 – 4.0	2.5 – 3.5
	Subbase	Expressway and Class I road	2.5 – 3.5	2.0 – 3.0	1.5 – 2.5
		Class II and lower class roads	2.0 – 3.0	1.5 – 2.5	1.0 – 2.0
Lime and fly ash stabilization	Base	Expressway and Class I road	≥1.1	≥1.0	≥0.9
		Class II and lower class roads	≥0.9	≥0.8	≥0.7
	Subbase	Expressway and Class I road	≥0.8	≥0.7	≥0.6
		Class II and lower class roads	≥0.7	≥0.6	≥0.5

Continued

Materials	Courses	Road classes	Super and very high traffic volume	High traffic volume	Medium and low traffic volume
Lime stabilization	Base	Class Ⅱ and lower class roads	≥0.8①		
	Subbase	Expressway and Class Ⅰ road	≥0.8		
		Class Ⅱ and lower class roads	0.5−0.7②		

Note: ① For soils with a plasticity index less than 7, the 7-day unconfined compressive strength of lime stabilized gravel and crushed stone should be more than 0.5 MPa.

② The lower limit is used for clay with a plasticity index less than 7, and the upper limit is used for clay with a plasticity index no less than 7.

3) Influencing factors

(1) Temperature

The chemical reactions inside the IBSM become more rapid and intense with the increase in temperature, and the strength increases accordingly. When the temperature is between 0−5 ℃, the strength of IBSM is difficult to be formed or has almost no strength gain. When the temperature is lower than 0 ℃, its strength may keep reducing if IBSM is subjected to repeated freezing and thawing. The IBSM may suffer more damage in the case of free water infiltrating. As a result, the IBSB should be constructed at a temperature higher than 5℃, and the construction should be ceased half a month (for the CSM) to one month (for the LSM) before the first freeze (−5 to −3 ℃).

(2) Age

The IBSM takes a considerable time to complete the whole chemical reaction process. Even for the CSM with high early strength, the hardening of the mix often lasts for more than one year after the cement finally sets. When the temperature is approximately constant, the strength and stiffness such as resilient modulus or elastic modulus of the IBSM grow with the increase of age. In particular, the hardening of LSM and LFASM is quite long.

The strength of the material is related not only to the material type, but also to the testing and curing conditions. According to specification JTG E51—2009, material composition is designed based on the 7-day unconfined compressive strength. Pavement design requires not only the compressive elastic modulus of the material, but also the tensile strength of the material, parameters of the material under standard testing conditions and field-testing conditions, the relationship between modulus and the change of time, etc. Generally, the design age of cement stabilized material or cement fly-ash stabilized material is 3 months, and the design age of lime or lime fly-ash stabilized material is 6 months.

5.3.2 Resilience Modulus Test

According to specifications AASHTO T 307—99 and JTG E51—2009 T0807/T0808, the resilience modulus test can be conducted to simulate the field confining stress and traffic loading on the base (Figure 5.7). The test uses the same specimen in the compression test with 100 mm or 150 mm in diameter. The test samples are prepared according to the determined optimum moisture content and cured. For CSM, the curing time is 90 days, while that is 180 days for LSM. After curing, the specimen is soaked in water for 24 hours and then taken out for the resilience modulus test. The cyclic load is applied to the specimen and the resilient or recovered axial deformation after unloading is recorded. The elastic deformation of the specimen is recorded after 1-minute loading at each stage. Then, the residual deformation is recorded after unloading for 0.5 minutes. Finally, the resilient deformation of the specimen under each load is calculated. The resilience modulus is calculated as:

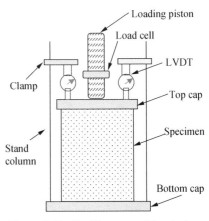

Figure 5.7 Resilience modulus test

$$M_r = \frac{\sigma_N L}{\varepsilon_r} \qquad (5.8)$$

where M_r = the resilience modulus (MPa);

σ_N = the cyclic (resilient) applied axial stress of the Nth load pulse (MPa);

ε_r = the resilient (recovered) axial deformation due to σ_N (mm);

L = the original specimen length (mm).

5.3.3 Split Tension Test

As specified in specification JTG E51—2009 T0806, the objective of the split tension test is to measure the tensile strength of IBSM. In the test, a cylinder with a height/diameter ratio of 1:1 is subjected to a compressive load at a constant rate applied on the sides until fracture, as shown in Figure 5.8. The radius of the arc surface is the same as the radius of the specimen, and its length should be greater than the height of the specimen. Table 5.3 shows the width of the strip and radius of the arc used by specimens of different sizes. Failure of the specimen occurs along its vertical diameter, due to tension developed in the transverse direction. The split or indirect tensile strength is calculated as Equations (5.9)–(5.13):

$$T = \frac{2P}{\pi Ld}\left(\sin 2\alpha - \frac{a}{d}\right) \quad (5.9)$$

where T = tensile strength (MPa);
  P = load at failure (N);
  L = length of specimen (mm);
  d = diameter of specimen (mm);
  α = central angle corresponding to half the width of the strip (°);
  a = width of strip (mm).

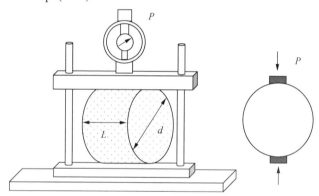

Figure 5.8 Split-tension test

Table 5.3 The size of the specimen with different specifications

Specimen size (mm)	Strip width (mm)	Arc radius (mm)
φ50 × 50	6.35	25
φ100 × 100	12.70	50
φ150 × 150	18.75	75

50-mm diameter

$$T = 0.012526 \frac{P}{L} \quad (5.10)$$

100-mm diameter

$$T = 0.006263 \frac{P}{L} \tag{5.11}$$

150-mm diameter

$$T = 0.004178 \frac{P}{L} \tag{5.12}$$

The static stiffness modulus of the stabilized material is calculated by Equation (5.13) based on the linear section of the load versus deformation curve obtained from the monotonic load split tension test.

$$E = P \times \frac{(v + 0.27)}{H \times H_c} \tag{5.13}$$

where E = static stiffness modulus (MPa);
P = load at failure (N);
v = Poisson's ratio;
H = recovered horizontal deformation of specimen after application of loads (mm);
H_c = height of specimen (mm).

5.3.4 Flexural Strength Test

According to specification JTG E51—2009 T0851, the flexural strength test is performed under the four point or third-point loading conditions. The specimen size depends on the particle size of the mix. For fine, medium, and coarse-grained soil, the dimension (width × height × span) of the beam is 50 mm × 50 mm × 200 mm, 100 mm × 100 mm × 400 mm, and 150 mm × 150 mm × 550 mm, respectively. The specimen is turned on its side and centered in the third-point loading apparatus, as illustrated in Figure 5.9. The load is continuously applied at a specified rate until rupture. If the fracture initiates in the middle third of the span length, the flexural strength is calculated as:

$$R_a = \frac{PL}{b^2 h} \tag{5.14}$$

where R_a = flexural strength (MPa);
b = width of specimen (mm);
h = height of specimen (mm);
L = span or the distance between two supports (mm);
P = ultimate loading at failure (N).

Note that the third-point loading ensures a constant bending moment without any shear force applied in the middle third of the specimen. Thus, Equation (5.14) is valid as long as fracture occurs in the middle third of the specimen. If the fracture occurs slightly outside the middle third, the results can still be used with some corrections. Otherwise, the results are discarded.

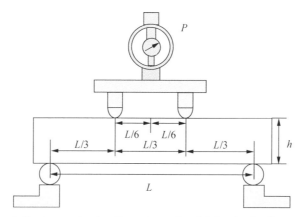

Figure 5.9 Apparatus for flexural test by third-point loading method

5.3.5 Fatigue Performance

The main failure type of pavement base under repeated traffic loading is a bending-tensile failure, and the design of pavement structure is mainly controlled by the flexural fatigue strength of materials. The indirect tensile fatigue test, the semi-circle bending fatigue test, and the beam fatigue test can all be used to evaluate the fatigue performance of IBSM. The specification JTG E51—2009 T0856 provides a beam fatigue test for IBSM using the same beam specimen and loading apparatus. It applies a repeated haversine load instead of a monotonic load. The beam specimen should be chosen based on the particle size and the flexural strength should be tested first. The fatigue life N_f can be obtained by a series of repeated loading according to the selected stress ratio (σ_f/σ_s), and then the fatigue equation can be established by regression.

The fatigue life of IBSM mainly depends on the type of materials and the stress ratio (σ_f/σ_s). Generally, if the stress ratio is less than 0.5, it is assumed that the material has an infinite fatigue life. Thus, the stress ratio is preferably set to be greater than 0.5 in the test. The fatigue performance can be illustrated through a scatter plot by the stress ratio and loading times (N_f). The relationship between the stress ratio and loading times is usually expressed as a double logarithmic fatigue equation $\lg N_f = a + b \lg \frac{\sigma_f}{\sigma_s}$ or a single logarithmic fatigue equation $\lg N_f = a + b\dfrac{\sigma_f}{\sigma_s}$. The fatigue equations of some IBSMs from previous studies are as follows:

(1) Cement stabilized macadam

50% probability

$$\lg N_f = 18.1574 - 18.1073\dfrac{\sigma_f}{\sigma_s} \tag{5.15}$$

95% probability

$$\lg N_f = 16.4642 - 18.1073\dfrac{\sigma_f}{\sigma_s} \tag{5.16}$$

(2) Lime and fly ash stabilized gravel

50% probability

$$\lg N_f = 15.5938 - 14.3697\dfrac{\sigma_f}{\sigma_s} \tag{5.17}$$

95% probability

$$\lg N_f = 14.5996 - 14.0631\dfrac{\sigma_f}{\sigma_s} \tag{5.18}$$

(3) Lime stabilized soil

50% probability

$$\lg N_f = 16.114 - 14.1\dfrac{\sigma_f}{\sigma_s} \tag{5.19}$$

95% probability

$$\lg N_f = 14.254 - 14.1\dfrac{\sigma_f}{\sigma_s} \tag{5.20}$$

(4) Cement stabilized Soil

50% probability

$$\lg N_f = 12.7972 - 11.2747\dfrac{\sigma_f}{\sigma_s} \tag{5.21}$$

95% probability

$$\lg N_f = 12.2287 - 11.2747\dfrac{\sigma_f}{\sigma_s} \tag{5.22}$$

(5) Lime and fly ash stabilized soil

50% probability

$$\lg N_f = 7.1069 - 4.493\dfrac{\sigma_f}{\sigma_s} \tag{5.23}$$

95% probability

$$\lg N_f = 6.2502 - 4.493 \frac{\sigma_f}{\sigma_s} \tag{5.24}$$

The equations above indicate that the fatigue life of the material under specific stress conditions depends on the strength and stiffness of the material (Figure 5.10). Materials with higher strength and smaller stiffness have longer fatigue lives. The fatigue equations vary under different confidence intervals due to the test variability. Small slopes in the fatigue equations indicate that a small change in the stress level has a greater impact on the fatigue life. High intercepts in the fatigue equations usually indicate the materials can withstand more load repetitions under the same stress level.

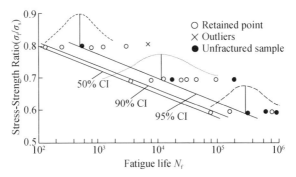

Figure 5.10 Fatigue curves of CSM with different confidence intervals (CI)

5.4 Shrinkage

Shrinkage in ISBM causes cracking and is the major concern for ISBM. The shrinkage in ISBM is caused by many reasons including cement hydration, drying, and temperature change. The greatest amount of shrinkage occurs early in the life of the pavement within the first couple of months. IBSB is usually built in warm seasons and inevitably experiences the evaporation of water before the sealing of the asphalt surface, which then causes the dry shrinkage from the base surface to the bottom. The daily temperature change also has an impact. Therefore, the IBSB at the early stage of construction is subject to both drying and thermal shrinkage and the drying shrinkage is primary at this stage. After the sealing of the asphalt layer, the humidity maintains approximately constant and the thermal shrinkage plays a dominant role at this stage. However, because of the insulating effect of the asphalt surface layer, the temperature change and the thermal shrinkage of the base are also reduced. Some studies have found that wide cracks (more than 6 mm) in CSM are primarily due to drying shrinkage.

5.4.1 Tests of Shrinkage

According to specification JTG E51—2009 T0854, the drying shrinkage properties are tested by measuring the length change and water loss of ISBM beams conditioned in a drying chamber with a constant temperature of 20 ℃. According to specification JTG E51—2009 T0855, the thermal shrinkage properties are tested by measuring the length and the temperature change of ISBM beams in a temperature chamber with constant humidity using a strain gauge. Several shrinkage parameters can be obtained based on the results of the two tests.

The drying or thermal shrinkage strain is the shrinkage per unit length of the beam specimen due to water loss or thermal change. The drying shrinkage strain is calculated by Equation (5.25), the thermal shrinkage strain is usually measured by the strain gauge.

$$\varepsilon_d = \frac{\Delta l}{l} \tag{5.25}$$

where ε_d = drying shrinkage strain ($\times 10^{-6}$)
Δl = shrinkage corresponding to water loss or temperature change (cm);
l = length of the specimen (cm).

The rate of water loss is the water loss per unit weight of the specimen.

$$\alpha_w = \frac{\Delta w}{w} \tag{5.26}$$

where α_w = rate of water loss (%);
Δw = weight of water loss (g);
w = weight of the specimen (g).

Drying shrinkage coefficient α_d ($\times 10^{-6}/\%$) is the drying shrinkage per unit rate of water loss

$$\alpha_d = \frac{\varepsilon_d}{\alpha_w} \tag{5.27}$$

The average drying shrinkage coefficient $\bar{\alpha}_d$ ($\times 10^{-6}/g$) is the drying shrinkage per unit water loss.

$$\bar{\alpha}_d = \frac{\varepsilon_d}{\Delta w} \tag{5.28}$$

The average thermal shrinkage coefficient $\bar{\alpha}_t$ ($\times 10^{-6}/°C$) is the thermal shrinkage per unit temperature change.

$$\bar{\alpha}_t = \frac{\varepsilon_t}{\Delta t} \tag{5.29}$$

where ε_t = thermal shrinkage strain ($\times 10^{-6}$)

Δt = temperature change (°C).

5.4.2 Drying Shrinkage Mechanism

Drying shrinkage of IBSM refers to the decrease of volume due to water loss. Similar to cement concrete, after mixing and compaction, water in the IBSM decreases continuously due to evaporation, hydration, and pozzolanic reaction. The removal of water in the form of capillary water, adsorbed water, and interlayer water causes the shrinkage of IBSM. The carbonation of $Ca(OH)_2$ also causes shrinkage since the volume of $CaCO_3$ is less than that of $Ca(OH)_2$. The drying shrinkage accounts for about 17% of the total shrinkage and is influenced by the content of the binder, the content of fine-grained soil (less than 0.5 mm), the plasticity index of the soil, the content of fine clay particles (less than 0.002 mm), minerals, water content, curing age, etc. The shrinkage mechanisms due to different types of water loss are as below.

(1) Capillarity Water

Due to the random growth of crystals and gels, two types of pores are left after hydration and pozzolanic reaction, including the interlayer hydration space and capillary voids. Free water surfaces in the capillary and crystal or gel pores will be in surface tension and cause shrinkage. It is noted that only the water in pores smaller than about 50 nm is subject to capillary tension force.

(2) Adsorbed Water

The adsorbed water is close to the solid surfaces with a maximum total thickness of about 1.3 nm under the influence of surface attractive force. After the loss of capillary water, the adsorbed water in the mix begins to lose as the relative humidity continues to decrease. The water film on the soil particles becomes thinner, the particle space becomes smaller, and the intermolecular force increases, which leads to additional shrinkage. The amount of shrinkage at this stage is greater than that of the capillary effect.

(3) Interlayer Water

The hydration gel pores between 0.5 nm and 2.5 nm are too small to influence the strength, but contribute 28% to the porosity of the paste. Interlayer water existing in those pores is under the influence of attractive force from two surfaces, and will therefore be more strongly held. At elevated temperatures and/or strong dry conditions such as less than 10% relative humidity, the interlayer water is removed and the van der Waals force pulls the surfaces together, causing considerable shrinkage.

5.4.3 Thermal Shrinkage Mechanism

Thermal shrinkage of IBSM refers to the decrease of volume due to a drop in temperature. The IBSM is a three-phase system consisting of solid particles, liquid, and gas. The thermal shrinkage of the IBSM is mainly determined by the thermal coefficient of the solid particles and water.

1) Shrinkage of Solid Phase

Most of the solid particles in the IBSM are crystals and some are amorphous. The thermal properties of the solid particles are determined by the composition, and the bonding and thermal motion between the particles. The bonding between the crystals is generally strong. The thermal motion of particles rises with the temperature and can only oscillate near the equilibrium position. Figure 5.11 presents the potential energy curve which is the energy of a molecule as a function of inter-atomic distance. At a specific temperature, the crystal particle oscillates between r' and r'' with an average spacing of $r = (r' + r'')/2$. With the increase of temperature, the particle tends to move a greater distance, and the average space r_0 shifts to the right.

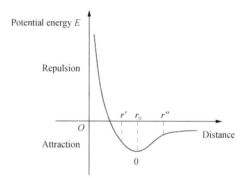

Figure 5.11 Potential energy curve of the crystal

Factors influencing the thermal expansion and shrinkage of crystals include the bonding force between particles, ionic charge, the distance of the particles, types of crystals, and the lattice structure. If the bonding force between particles is stronger, the ions get closer, the ionic charge gets larger, and the thermal expansion coefficient reduces. The larger the coordination number of the crystal particle, the larger the thermal expansion-shrinkage coefficient. For the layered crystals, the thermal expansion coefficient perpendicular to the layer is larger than that parallel to the layer. The thermal expansion coefficients of dense-packed structures is larger than those of open structures.

In the IBSM, the thermal expansion coefficients of the raw materials are generally lower. The clay has a greater thermal expansion coefficient, while fly ash has the least value. The thermal expansion coefficients of gel products from hydration and pozzolanic reaction are generally higher. The thermal expansion coefficients of the IBSM are the results of all of its solid components.

2) Influence of water

The large voids, capillary pores, and interlayer hydration gel pores in the IBSM contain free water, capillary water, and interlay water, respectively. Water has a great influence on the thermal shrinkage of the IBSM through expansion action, capillary action, and freezing action.

Water has a relatively high thermal expansion coefficient of $70 \times 10^{-6}/°C$, which is $4-7$ times of the solid phase. The expansion pressure of water increases the distance between particles and causes expansion when the temperature rises. Capillary tension force exists only when the water content is within a certain range. If the material is too dry or too wet, the capillary tension force disappears. Therefore, the thermal shrinkage coefficient of the material in the dry and saturated state should be lower than that in the unsaturated state. If the water in the voids freezes below freezing, the volume increases by 9%, causing expansion.

5.5　Mix Design

The mix design of IBSM mainly includes three parts: target mix design, production mix design, and determination of construction parameters.

(1) The target mix design includes the selection of gradation, determination of binder and content, and verification of design or construction requirements.

(2) The production mix design includes the determination of plant ratio, the allowable handling time of CSM, the control curve of binder content, the optimal water content, and the maximum dry density of the mix.

(3) The determination of construction parameters includes the determination of binder content, water content, and maximum dry density during construction, and verification of the strength of the mix.

Civil Engineering Materials

The 7-day compressive strength is used as the major design criteria for the mix design of IBSM. The 7-day compressive strength should meet the requirements in Table 5.2.

5.5.1 Design Procedure

The procedure of mix design are shown in Figure 5.12 and the details are discussed below.

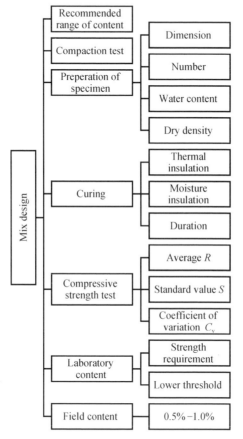

Figure 5.12 Mix Design Procedure

1. Selecting representative samples of raw materials.

2. Determining the range of binder content as recommended in Table 5.4. The content of the binder is the percentage of the weight of the binder to the total weight of the mix. At least three groups of mix with the minimum, average, and maximum binder content are required. The other two kinds of binder content can be determined by interpolation. Table 5.5 and Table 5.6 show the recommended proportions of CIWSM and LIWSM.

5 Inorganic Binder Stabilized Materials

Table 5.4 Recommended binder content for CSM

Materials	Pavement layers	Types	Recommended content (%)
Crushed gravel and stone	Base	$R_d \geq 5$ MPa	5,6,7,8,9
Crushed gravel and stone	Base	$R_d < 5$ MPa	3,4,5,6,7
Soils, sand, and stone chips	Base	Plasticity index < 12	5,7,9,11,13
Soils, sand, and stone chips	Base	Plasticity index ≥ 12	8,10,12,14,16
Crushed gravel and stone	Subbase	—	3,4,5,6,7
Soils, sand, and stone chips	Subbase	Plasticity index < 12	4,5,6,7,8
Soils, sand, and stone chips	Subbase	Plasticity index ≥ 12	6,8,10,12,14
RCC lean concrete	Base	—	7,8.5,10,11.5,13

Table 5.5 Recommended proportions of CIWSM

Stabilizers	Materials	Pavement Layers	Proportion of binder	Ratio of binder over soils
Cement and fly ash	Cement and silica aluminum fly ash[1]	Base or subbase	Cement: fly ash = 1:9–1:3	—
Cement and fly ash	Cement and fly ash stabilized soils	Base or subbase	Cement: fly ash = 1:5–1:3	Cement and fly ash: soils = 10:90 – 30:70[2]
Cement and fly ash	Cement and fly ash stabilized graded macadam or gravel	Base	Cement: fly ash = 1:5–1:3	Cement and fly ash: macadam or gravel = 15:85[3] – 20:80
Cement and cinder	Cement and cinder	Base or subbase	Cement: cinder = 5:95 – 15:85	—
Cement and cinder	Cement and cinder stabilized soils	Base or subbase	Cement: cinder = 1:2–1:5	Cement and cinder: soils = 1:2–1:5[4]
Cement and cinder	Cement and cinder stabilized material	Base or subbase	Cement: cinder: soil = (3–5):(26–33):(71–62)	

Notes: [1] It is a silica aluminum fly ash with a CaO content of 2%–6%.

[2] The ratio of cement to fly ash should be 1:3–1:2 when this proportion is adopted.

[3] When the ratio of cement and fly ash to aggregates is 20:80 – 15:85, the aggregates form a skeleton, and the cement and fly ash can fill the voids and cement the mix.

[4] Cement in the mix should be more than 4% and mix proportion of high strength can be selected through tests.

Table 5.6 Recommended proportions of LIWSM

Stabilizers	Materials	Pavement Layers	Proportion of binder	Ratio of binder over soils
Lime and fly ash	Lime and silica aluminum fly ash[①]	Base or subbase	Lime: fly ash = 1:9–1:2	—
	Lime and fly ash stabilized soils	Base or subbase	Lime: fly ash = 1:4[②]–1:2	Lime and fly ash: soils = 10:90 – 30:70[③]
	Lime and fly ash stabilized graded macadam or gravel	Base	Lime: fly ash = 1:4–1:2	Lime and fly ash: macadam or gravel = 15:85[④]–20:80
Lime and cinder	Lime and cinder	Base or subbase	Lime: cinder = 15:85 – 20:80	—
	Lime and cinder stabilized soils	Base or subbase	Lime: cinder = 1:4–1:1	Lime and cinder: fine-grained materials = 1:4[⑤]–1:1
	Lime and cinder stabilized material	Base or subbase	Lime: cinder: soils = (7–9):(26–33):(67–58)	

Notes: ① It is a silica aluminum fly ash with 2%–6% CaO content.

② 1:2 is proper for silt.

③ The ratio of lime to fly ash should be 1:3–1:2 when this proportion is adopted.

④ When the ratio of lime and fly ash to granular materials is 15:85 – 20:80, the aggregates form the skeleton, and lime and fly ash can fill the voids and bond the mix. This mix is called dense skeleton lime-fly ash-stabilized aggregates.

⑤ The lime in the mix should be more than 10% and the mix proportion of high strength can be selected through tests.

3. Determining the optimum water content and maximum dry density of the soils through compaction tests according to specification JTG E51—2009 T0804, as shown in Figure 5.13.

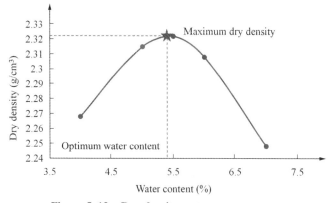

Figure 5.13 Dry density-water content curve

5 Inorganic Binder Stabilized Materials

4. Preparing specimens according to the optimum water content and calculating the dry density considering the specified degree of compaction in the field.

5. Curing the specimens for 6 days at the specified temperature (20 ℃ ±2 ℃ in the northern region and 25 ℃ ±2 ℃ in the southern region) and then immerse them in water for 1 day before the strength test.

The number of specimens for each group should meet the requirements in Table 5.7 when performing the strength test. If the coefficient of variation of the results of the strength test is greater than the requirements in Table 5.7, the test should be repeated to find out the causes, and measures should be taken. If the coefficient of variation cannot be reduced, the number of specimens should be increased. For coarse-grained soils, if the coefficients of variation of multiple tests are all less than 20%, nine specimens are enough. For both the coarse and medium-grained soils, if the coefficients of variation of multiple tests are all less than 15%, six specimens are sufficient.

Table 5.7 Minimum number of specimens for one group in the strength test

Materials	Requirements of coefficient of variation		
	<10%	10%-15%	15%-20%
Fine-grained soils[①]	6	9	—
Medium-grained soils[②]	6	9	13
Coarse-grained soils[③]	—	9	13

Notes: ① Soils with a nominal maximum particle size less than 16 mm.
② Soils with a nominal maximum particle size between 16 mm and 26.5 mm.
③ Soils with a nominal maximum particle size greater than 26.5 mm.

6. Selecting the binder content according to the strength requirements of the mix. The average compressive strength $\bar{R}_7$ should meet the requirements of Equation (5.7).

7. Considering the difference between the laboratory tests and the field conditions, the actual content of the binder used in the field should be 0.5%-1.0% more than the lab results. Usually, 0.5% is recommended for a good mixing effort and 1% is recommended for a poor mixing effort.

5.5.2 Requirements of Materials

Details of the requirements of materials including cement, lime, fly ash, soils, and gradations are discussed below. All of the requirements are from the specification JTG/T F20—2015.

1) Cement stabilized material

(1) Soil

Any fine, medium and coarse-grained soils that can be crushed can be stabilized with cement, including stone ballast, chips, gravel, crushed stone, gravelly soil, etc. Coarse aggregates should meet the Grade I requirements in Table 5.8.

Table 5.8 Quality requirements of coarse aggregates used in IBSM (%)

Test items	Pavement layers	Expressways and Class I roads				Class II and lower class roads		Test methods
		Super and very high traffic volume		High, medium, and light traffic volume				
	Grades	I	II	I	II	I	II	
Max. crushing value	Base	22	22	26	26	35	30	T 0316
	Subbase	30	26	30	26	40	35	
Max. content of the flaky and elongated particle	Base	18	18	22	18	—	20	T 0312
	Subbase	—	20	—	20	—	20	
Max. dust (<0.075 mm) content	Base	1.2	1.2	2	2	—		T 0310
	Subbase	—	—	—	—	—	—	
Max. soft particle content	Base	3	3	5	5			T 0320
	Subbase	—	—	—	—	—	—	

Table 5.9 lists the types of coarse aggregates in CSM. The nominal maximum size of graded crushed stone and gravel should not be greater than 26.5 mm for the base of expressways or Class I roads, 31.5 mm for the base of Class II or lower class roads, or 37.5 mm for the subbase layer.

Table 5.9 Different types of coarse aggregates

Types	Engineering size (mm)	Passing (%)									Nominal maximum size (mm)
		53	37.5	31.5	26.5	19.0	13.2	9.5	4.75	2.36	
G1	20 – 40	100	90 – 100	—	—	0 – 10	0 – 5	—	—	—	19 – 37.5
G2	20 – 30	—	100	90 – 100	—	0 – 10	0 – 5	—	—	—	19 – 31.5

5 Inorganic Binder Stabilized Materials

Continued

Types	Engineering sizes (mm)	Passing (%)									Nominal maximum sizes (mm)
		53	37.5	31.5	26.5	19.0	13.2	9.5	4.75	2.36	
G3	20 – 25	—	—	100	90 – 100	0 – 10	0 – 5	—	—	—	19 – 26.5
G4	15 – 25	—	—	100	90 – 100	—	0 – 10	0 – 5	—	—	13.2 – 26.5
G5	15 – 20	—	—	—	100	90 – 100	0 – 10	0 – 5	—	—	13.2 – 19
G6	10 – 30	—	100	90 – 100	—	—	—	0 – 10	0 – 5	—	9.5 – 31.5
G7	10 – 25	—	—	100	90 – 100	—	—	0 – 10	0 – 5	—	9.5 – 26.5
G8	10 – 20	—	—	—	100	90 – 100	—	0 – 10	0 – 5	—	9.5 – 19
G9	10 – 15	—	—	—	—	100	90 – 100	0 – 10	0 – 5	—	9.5 – 13.2
G10	5 – 15	—	—	—	—	100	90 – 100	40 – 70	0 – 10	0 – 5	4.75 – 13.2
G11	5 – 10	—	—	—	—	—	100	90 – 100	0 – 10	0 – 5	4.75 – 9.5

The fine aggregates for the base of the expressways and Class I roads should meet the requirements in Table 5.10.

Table 5.10 Quality requirements of fine aggregates used in IBSB

Test items	CSM	LSM	LFASM	CFASM	Test methods
Plasticity indexes	≤ 17	15 – 20	12 – 20	—	T 0118
Max. organic content (%)	<2	10	10	<2	T 0313
Max. sulfate content (%)	0.25	0.8	—	0.25	T 0341

Table 5.11 lists the types of fine aggregates used in CSM. For fine aggregates of 0 – 3 mm and 0 – 5 mm, the content of particles larger than 2.36 mm and 4.75 mm should be strictly controlled. For fine aggregates of 3 – 5 mm, the content of <2.36 mm particles should be strictly controlled. For expressways and Class I roads, the content of <0.075 mm particles should be no more than 15%. For Class II or lower class roads, the content of <0.075 mm particles should be no more than 20%.

Table 5.11 Different types of fine aggregates

Gradations	Engineering sizes (mm)	Passing (%)							Nominal maximum sizes (mm)
		9.5	4.75	2.36	1.18	0.6	0.3	0.15	0.075	
XG1	3 – 5	100	90 – 100	0 – 15	0 – 5	—	—	—	—	2.36 – 4.75
XG2	0 – 3	—	100	90 – 100	—	—	—	0 – 15	0 – 2.36
XG3	0 – 5	100	90 – 100	—	—	—	—	0 – 20	0 – 4.75

Table 5.12 lists the recommended gradations for CSM. Table 5.13 lists the recommended gradations for CFASM.

Table 5.12 Recommended gradations of CSM (%)

Sieve sizes (mm)	Expressways and Class I roads			Class II and lower class roads		
	C-B-1	C-B-2	C-B-3	C-C-1	C-C-2	C-C-3
37.5	—	—	—	100	—	—
31.5	—	—	100	100 – 90	100	—
26.5	100	—	—	94 – 81	100 – 90	100
19	86 – 82	100	68 – 86	83 – 67	87 – 73	100 – 99
16	79 – 73	93 – 88	—	78 – 61	82 – 65	92 – 79
13.2	72 – 65	86 – 76	—	73 – 54	75 – 58	83 – 67
9.5	62 – 53	72 – 59	38 – 58	64 – 45	66 – 47	71 – 52
4.75	45 – 35	45 – 35	22 – 32	50 – 30	50 – 30	50 – 30
2.36	31 – 22	31 – 22	16 – 28	36 – 19	36 – 19	36 – 19
1.18	22 – 13	22 – 13	—	26 – 12	26 – 12	26 – 12
0.6	15 – 8	15 – 8	8 – 15	19 – 8	19 – 8	19 – 8
0.3	10 – 5	10 – 5	—	14 – 5	14 – 5	14 – 5
0.15	7 – 3	7 – 3	—	10 – 3	10 – 3	10 – 3
0.075	5 – 2	5 – 2	0 – 3	7 – 2	7 – 2	7 – 2

Table 5.13 Recommended gradations of CFASM   (%)

Sieve sizes (mm)	Expressways and Class I roads				Class II and lower class roads			
	Stabilized macadam		Stabilized gravel		Stabilized macadam		Stabilized gravel	
	CF-A-1S	CF-A-2S	CF-A-1L	CF-A-2L	CF-B-1S	CF-B-2S	CF-B-1L	CF-B-2L
37.5	—	—	—	—	100	—	100	—
31.5	100	—	100	—	100–90	100	100–90	100
26.5	95–90	100	95–91	100	93–80	100–90	94–81	100–90
19	84–72	88–79	85–76	89–82	81–64	86–70	83–67	87–73
16	79–65	82–70	80–69	84–73	75–57	79–62	78–61	82–65
13.2	72–57	76–61	75–62	78–65	69–50	72–54	73–54	75–58
9.5	62–47	64–49	65–51	67–53	60–40	62–42	64–45	66–47
4.75	40–30	40–30	45–35	45–35	45–25	45–25	50–30	50–30
2.36	28–19	28–19	33–22	33–22	31–16	31–16	36–19	36–19
1.18	20–12	20–12	24–13	24–13	22–11	22–11	26–12	26–12
0.6	14–8	14–8	18–8	18–8	15–7	15–7	19–8	19–8
0.3	10–5	10–5	13–5	13–5	—	—	—	—
0.15	7–3	7–3	10–3	10–3	—	—	—	—
0.075	5–2	5–2	7–2	7–2	5–2	5–2	7–2	7–2

(2) Cement

Ordinary Portland cement, slag Portland cement, and pozzolanic Portland cement can be used for stabilizing soils. Considering that it takes a specific time to mix and transport before compaction, the cement with a long final setting time (more than 6 h) should be selected. It is advisable to use cement with a lower strength grade, such as grade 32.5 or 42.5.

(3) Fly ash

Fly ash can be added to the CSM and should meet the requirements in Table 5.14. Coal gangue, cinder, blast furnace slag, steel slag, and other industrial wastes, etc. can also be added. The strength, drying shrinkage, and thermal shrinkage at different ages and temperatures should be evaluated during mix design. When used as aggregates, the nominal maximum size of industrial wastes should be less than 31.5 mm, and should also have a specific gradation.

Table 5.14 Requirements of fly ash used in CFASM

Test items	Requirements (mg/L)	Test methods
Min. SiO_2, Al_2O_3 and Fe_2O_3 content (%)	> 70	T 0816
Max. ignition loss (%)	20	T 0817
Min. specific surface area (cm^2/g)	> 2500	T 0820
Min. 0.3 mm sieve passing percentage (%)	90	T 0818
Min. 0.075 mm sieve passing percentage (%)	70	T 0818
Max. moisture content (%)	35	T 0801

(4) Water

Water used for CSM should meet the requirements in Table 5.15.

Table 5.15 Requirements of water used in CSM

Test items	Requirements (mg/L)	Test methods
Min. Cl^- content	4.5	
Max. SO_4^{2-} content	3500	
Max. alkali content	2700	JGJ 63
Max. content of soluble solid	10,000	
Max. content of insoluble solid	5000	

2) Lime stabilized material

Any soils that can be crushed can be stabilized with lime, mainly fine-grained soils. According to specification JTG/T F20—2015, the lime used in the LSM should meet the requirements in Table 5.16 and Table 5.17.

Table 5.16 Requirements of quicklime used in LSM (%)

Test items	Calcitic quicklime			Dolomite quicklime			Test methods
	I	II	III	I	II	III	
Min. CaO and MgO content	85	80	70	80	75	65	T 0813
Max. impurity content	7	11	17	10	14	20	T 0815
MgO content	≤5			>5			T 0812

Table 5.17   Requirements of slaked lime used in LSM   (%)

Test items		Calcitic slaked lime			Dolomite slaked lime			Test methods
		I	II	III	I	II	III	
Min. CaO and MgO content		65	60	55	60	55	50	T 0813
Max. water content		4	4	4	4	4	4	T 0801
Fineness	Max. retained on 0.6 mm sieve	0	1	1	0	1	1	T 0814
	Max. retained on 0.15 mm sieve	13	20	—	13	20	—	T 0814
MgO content		≤4			>4			T 0812

3) Lime and fly ash stabilized material

The quality of lime in the LFASM should be higher than the Grade III's requirements listed in Table 5.16 and Table 5.17. The fly ash should meet the requirements in Table 5.14. The requirements of coarse and fine aggregates are shown in Table 5.8 and Table 5.10, respectively. The recommended gradation of LFASM is listed in Table 5.18.

Table 5.18   Recommended gradation of LFASM   (%)

Sieve sizes (mm)	Expressways and Class I roads				Class II and lower class roads			
	Stabilized macadam		Stabilized gravel		Stabilized macadam		Stabilized gravel	
	LF-A-1S	LF-A-2S	LF-A-1L	LF-A-2L	LF-B-1S	LF-B-2S	LF-B-1L	LF-B-2L
37.5	—	—	—	—	100	—	100	—
31.5	100	—	100	—	100−90	100	100−90	100
26.5	95−91	100	96−93	100	94−81	100−90	95−84	100−90
19	85−76	89−82	88−81	91−86	83−67	87−73	87−72	91−77
16	80−69	84−73	84−75	87−79	78−61	82−65	83−67	86−71
13.2	75−62	78−65	79−69	82−72	73−54	75−58	79−62	81−65
9.5	65−51	67−53	71−60	73−62	64−45	66−47	72−54	74−55
4.75	45−35	45−35	55−45	55−45	50−30	50−30	60−40	60−40
2.36	31−22	31−22	39−27	39−27	36−19	36−19	44−24	44−24
1.18	22−13	22−13	28−16	28−16	26−12	26−12	33−15	33−15
0.6	15−8	15−8	20−10	20−10	19−8	19−8	25−9	25−9
0.3	10−5	10−5	14−6	14−6	—	—	—	—
0.15	7−3	7−3	10−3	10−3	—	—	—	—
0.075	5−2	5−2	7−2	7−2	7−2	7−2	10−2	10−2

5.5.3 Requirements of Compaction

The compaction of the IBSM base and subbase during construction should meet the requirements in Table 5.19. For graded gravel materials, the compaction of the base should not be less than 99%, and that of the subbase should not be less than 97%. The compaction of the base and subbase of the expressways and Class I roads can be increased by 1%–2% under the super and very high traffic volume.

Table 5.19 Minimum compaction requirements of IBSB (%)

Pavement layers	Road classes	Soils	CSM	LSM	LFASM	CFASM
Base	Expressways and Class I roads	—	—	98	—	98	98
Base	Class II and lower class roads	Coarse and medium-grained soils	97	97	97	97
Base	Class II and lower class roads	Fine-grained soils	95	95	95	95
Subbase	Expressways and Class I roads	Coarse and medium-grained soils	97	97	97	97
Subbase	Expressways and Class I roads	Fine-grained soils	95	95	95	95
Subbase	Class II and lower class roads	Coarse and medium-grained soils	95	95	95	95
Subbase	Class II and lower class roads	Fine-grained soils	93	93	93	93

Questions

1. What are the advantages and disadvantages of IBSM?
2. How the strength is developed in the CSM and LFASM?
3. What is the difference between the two classes of fly ash?
4. What chemical compounds contribute to early strength gain for IBSM?
5. What kind of strength tests are usually conducted to evaluate the IBSM?
6. How to evaluate the fatigue performance of IBSM?
7. How to test the shrinkage of IBSM?
8. What are the main causes of the drying shrinkage of IBSM?
9. Discuss the mix design procedure of IBSM.
10. Why is water very important for both drying and thermal shrinkage of IBSM?

6 Asphalt

The word "asphalt" is usually used in the US, while Europe uses the word "bitumen" more often. Asphalt is used mostly in pavement construction. It is also used as a sealing and waterproofing agent for roofs and foundations. Human has a very long history of using asphalt to pave roads. Around 5000 years ago, ancient Babylonians built an asphalt block road. In the 1830s, the British started to build the tarmacadam, which is to pour asphalt on an aggregate layer. The US built its first asphalt pavement with natural asphalt in the 1870s and the first pavement with petroleum asphalt in 1913. In 1925, China built its first asphalt pavement with petroleum asphalt on the Zhongshan Road in Nanjing. This chapter covers the classification, production, composition, structure, tests, properties, aging, modification, and the grading systems of asphalt.

6.1 Classification and Production

6.1.1 Classification

According to its different origins, asphalt can be divided into two categories: asphalt and tar. Asphalt can either be found in natural deposits or refined from petroleum. Natural asphalt refers to the product of petroleum under natural conditions, and it is formed by long-term geophysical actions. Trinidad lake asphalt and rock asphalt are two typical natural asphalt. Petroleum asphalt is the sticky, black, and highly viscous liquid or semi-solid residue during the refining of petroleum by steaming and other processes. Tar is a by-product or distillation of coke production from coal. Tar was used to pave roads, but is now primarily used for waterproofing membranes. Asphalt is soluble in petroleum products while tar is insoluble in a petroleum solvent.

6.1.2 Production

The quantity and quality of asphalt depend on the crude petroleum sources and the refining methods. Some crude sources produce little asphalt, while others have high asphalt contents. For nearly 1500 types of petroleum, those from the Middle East and South America produce good-quality asphalt. A petroleum refinery's job is to break crude oil hydrocarbon molecule chains down into slightly different groupings of molecules of molecular weight, through distillation with heat, chemical reactions, and changes in pressure. Different products including gas, gasoline, kerosene, diesel, lubricating oil, heavy gas oil, and asphalt are separated at different temperatures. Since asphalt is a less valuable product than other components of crude oil, refineries are set up to produce more valuable fuels such as gasoline, kerosene, and diesel oil, at the expense of asphalt production. The production methods of asphalt include:

1) Distillation

Crude oil is separated into gasoline, kerosene, diesel oil, and other components according to their different boiling points through the atmospheric distillation tower and vacuum distillation tower, and then the residue is obtained as asphalt. The one-stage and two-stage distilled asphalt fractions are referred to as straight-run asphalt.

2) Oxidation

The oxidation process involves the reaction and rearrangement of asphalt molecules in the presence of oxygen at elevated temperatures. In the oxidation tower, the straight-run asphalt is heated, the air is blown and sometimes the catalyst is added. The reactions include dehydrogenation, oxidation, and polycondensation. The low molecular weight hydrocarbons in the asphalt are transformed into high molecular weight hydrocarbons. This type of asphalt is mostly used for industrial applications such as roofing and pipe coatings, and low penetration grade pavement asphalt.

3) Semi-oxidation

The straight-run asphalt may be too soft and have high-temperature susceptibility. The over-oxidized asphalt may be too still and have poor low-temperature performance. Semi-oxidation is improved oxidation with lower oxidation temperature, longer time, and lower air volume to obtain the asphalt with better high- and low-temperature performance.

4) Solvent deasphalting

Solvent deasphalting is a pretreatment process for inferior residual oil. It uses heavy oil such as vacuum residue as raw materials, and extracts hydrocarbons such as propane and butane as solvents. The extract is deasphalted oil, which is usually used as a heavy lubricating oil raw material or a cracking raw material. The residue is the asphalt.

5) Blending

Asphalt with different viscosity or other properties can be blended to produce blended asphalt that meet the requirements. The proportions of asphalts with different viscosities can be determined through laboratory tests, calculation methods, or component adjustment methods according to the requirements.

6) Cutback

Cutback asphalt is produced by dissolving asphalt in a light molecular weight hydrocarbon solvent such as kerosene, gasoline. When the cutback is sprayed on pavements or mixed with aggregates, the solvent evaporates, leaving the asphalt residue as the binder. Depending on the evaporation rate of the solvent, we have slow, medium, and rapid curing cutbacks. In the past, cutbacks were widely used for pavement construction. However, the cost of expensive solvents increases, and cutbacks are hazardous materials due to the volatility of the solvents. Therefore, the use of cutbacks is limited and is mainly for pothole patching now.

7) Emulsification

An alternative to dissolving the asphalt in a solvent is to physically break asphalt down into micron-sized globs that are mixed into water containing an emulsifying agent. Emulsified asphalt typically consists of about 60% to 70% asphalt, 30% to 40% water, and a fraction of a percentage of

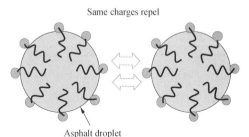

Figure 6.1 Asphalt emulsification

emulsifying agents. As shown in Figure 6.1, an emulsifying molecule has two distinct components, the head portion, which has an electrostatic charge, and the tail portion, which has a high affinity for asphalt. The same charges of emulsifying molecules repel, pushing asphalt globs apart, and force asphalt and water, two kinds of immiscible liquids, to combine into a suspension substance. When the water in asphalt emulsions evaporates, the asphalt globs are allowed to come together, and form the binder. The separation between the asphalt residue and water is referred to as breaking or setting. According to the type of emulsifier, emulsified asphalt can be divided into cationic emulsified asphalt, anionic emulsified asphalt, and non-ionic emulsified asphalt. Depending on the rate of emulsion setting, we also have slow, medium, and rapid setting asphalt emulsion.

8) Modification

Modified asphalt is the asphalt prepared by mixing rubber, resin, high polymer, natural asphalt, ground rubber powder, or other additives or modifiers. The typical polymer modifiers are either plastics or rubber and alter the strength and viscoelastic properties of the asphalt by increasing its elastic response, improving its cohesive and fracture strength, and providing greater ductility.

6.2 Composition and Structure

6.2.1 Composition

Asphalt contains various hydrocarbons and their nonmetallic derivatives (oxygen, sulfur, nitrogen). The percentages of the chemical components and the molecular structure, vary depending on the crude oil source. Its components are mainly carbon (80%–87%), hydrogen (10%–15%), and non-hydrocarbon elements, such as oxygen, sulfur, nitrogen (<3%), and very few metals. The constituents of asphalt are determined based on whether or not they are soluble in different solvents. It may be firstly divided into two main fractions: asphaltenes which are insoluble in a light aliphatic hydrocarbon solvent such as n-heptane and maltenes which are soluble in n-heptane. The maltenes can be further divided into resins which are highly polar hydrocarbons and oil which can be subdivided into aromatics and saturates.

1) Three fractions

The three fractions of asphalt include asphaltenes, resins, and oil. Table 6.1 summarizes the characteristics of the three fractions. The asphaltenes are responsible for the viscosity and the adhesive property of the asphalt. If the asphaltene content is less than 10%, the asphalt concrete will be difficult to compact to the proper construction density. Resins are dark and semisolid or solid, with a viscosity that is largely affected by temperature. The resins act as agents to disperse asphaltenes in the oil. The resin's viscosity is largely affected by temperature. When the resins are oxidized, for example during aging, they yield asphaltene-type molecules. Resins can be divided into neutral resin and acid resin. Neutral resin increases the asphalt's plasticity, fluidity, and bonding. Acidic resin is the most active component in asphalt. It can improve the infiltration of asphalt to mineral materials, especially improve the adhesion with the carbonate rock, and increase the emulsification of asphalt. Oil is a clear or white liquid and is soluble in most solvents, allowing asphalt to flow.

Table 6.1 Characteristics of the three fractions in asphalt

Fractions	Appearance	Molecular weight	Ratio of Carbon/hydrogen	Content (%)	Specific gravity	Physicochemical characteristics
Oil	Clear or white liquid	200 – 700	0.5 – 0.7	45 – 60	0.7 – 1.0	Almost soluble in most organic solvents and has optical activity
Resin	Brown gel or solid	800 – 3000	0.7 – 0.8	15 – 30	1.0 – 1.1	High-temperature susceptibility, melting point is less than 100 ℃
Asphaltene	Dark brown, friable solid	1000 – 5000	0.8 – 1.0	5 – 30	1.1 – 1.5	Does not melt but carbonized when heated

2) Four fractions

When dividing oil into aromatics and saturates, we have the four fractions of asphalt, including asphaltenes, resins, aromatics, and saturates, which account for approximately 15%, 20%, 50%, and 15% of the total. The aromatic oil is oily and yellow or brown in appearance and includes naphthenoaromatic type rings. The saturated oil is made up mainly of long straight saturated chains and appears as a highly viscous colorless oil. Table 6.2 summarizes the characteristics of the four fractions.

Table 6.2 Characteristics of the four fractions in asphalt

Fractions	Appearance	Specific gravity	Molecular weight	Chemical structures
Saturate	Colorless liquid	0.89	625	Alkane and cycloalkane
Aromatic	Yellow liquid	0.99	730	Aromatic hydrocarbon
Resin	Brown gel	1.09	970	Multiple-ring structure
Asphaltene	Dark brown solid	1.15	3400	Condensed structure

6.2.2 Colloidal Structure

Asphalt is a colloidal system consisting of high molecular weight asphaltene micelles dispersed or dissolved in a lower molecular weight oily medium or the maltenes (Figure 6.2). Asphaltenes can cluster together to form large particles. Various components of asphalt interact with each other to form a balanced or compatible system. This balance of components makes the asphalt suitable as a binder.

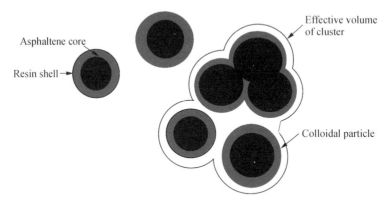

Figure 6.2　Asphalt colloidal structure

According to the different chemical composition and relative content of each component in asphalt, the colloidal structure of asphalt can be classified into three types, which are sol type, sol-gel type, and gel type (Figure 6.3). The sol type includes less than 10% asphaltenes, in which the asphaltene particles are complete dispersal. It is usually the refinery asphalt. It is soft at low temperatures and is highly temperature-dependent. The gel type includes higher than 30% asphaltenes in which the asphaltene particles are connected, forming a comb-shaped network. It is usually the oxidized asphalt. It is very brittle at low temperatures and its temperature dependency is low. Between the two types is the sol-gel type containing 15%–25% asphaltenes. Its low-temperature brittleness and temperature susceptibility are both moderate.

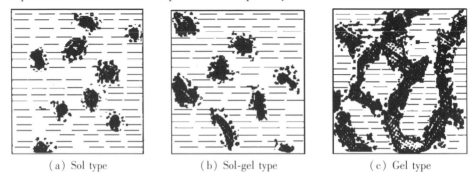

Figure 6.3　Colloidal structure of asphalt

6.3 Properties

6.3.1 Physical Properties

1) Density

The density of asphalt varies with the change in temperature. At 15−25 °C, the density of asphalt ranges from 0.96 to 1.04. The density of asphalt is a key parameter for asphalt mixture design. Preservation and transportation of asphalt also need to consider the density and thermal expansion of asphalt.

2) Coefficient of expansion

The change in volume of asphalt at a temperature rise of 1 °C is the coefficient of bulk expansion. The coefficient of expansion of asphalt can be determined based on the density at different temperatures as shown in Equation (6.1), it is the change of density over the change of temperature.

$$A = \frac{D_{T2} - D_{T1}}{D_{T1}} \times \frac{1}{T_1 - T_2} \quad (6.1)$$

where D_{T2} = density of asphalt at temperature T_2 (g/cm^3);

D_{T1} = density of asphalt at temperature T_1 (g/cm^3);

T_1, T_2 = two testing temperatures (°C).

6.3.2 Penetration

According to specification JTG E20—2011 T0604, penetration refers to the penetration depth of a 100 g needle into asphalt in 5 s at 25 °C (Figure 6.4). The depth of penetration, in units of 0.1 mm, is recorded and reported as the penetration value. The penetration is inversely related to viscosity, as the test measures the viscous resistance to the penetration of a needle into a container of asphalt. Penetration is the most widely used method to measure the relative viscosity or consistency of asphalt in the world and is still used to determine the asphalt grade in many countries.

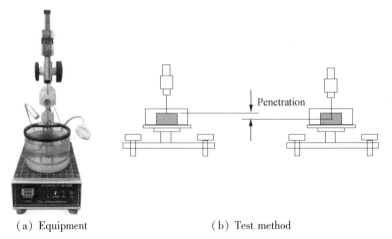

(a) Equipment   (b) Test method

Figure 6.4 Penetration test of asphalt

6.3.3 Viscosity

In 1930, Physicist Thomas Parnell set up the Pitch Drop Experiment (Figure 6.5). It is one of the longest-running lab tests. At room temperature, asphalt still exhibits viscosity over a very long period. The asphalt drops formed slowly and approximately fell a drop every ten years. Viscosity is a very important property of asphalt. It is a measure of the resistance of a material to deformation at a given rate. Viscosity can be measured directly in various ways. The penetration, softening point and different viscosity tests can all be used to evaluate the viscosity or consistency of asphalt. The different types of viscosities include relative viscosity, absolute or dynamic viscosity, and kinematic viscosity.

Figure 6.5 Pitch Drop Experiment

1) Relative viscosity

Relative viscosity can be measured by the Ford viscosity cup (Figure 6.6). It is also called the standard viscosity. Relative viscosity refers to the time that 50 mL asphalt needs to flow through a nozzle. Penetration and softening point tests are also relative viscosity tests.

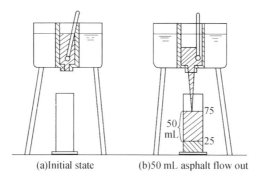

(a) Initial state　　　(b) 50 mL asphalt flow out

Figure 6.6　Relative viscosity test using the Ford viscosity cup

2) Absolute or dynamic viscosity

The absolute or dynamic viscosity can be measured by three different but similar types of equipment, including the capillary viscometer, the rotational viscometer, and the sliding plate viscometer.

(1) The capillary viscometer measures the time for asphalt to flow through a small tube under a controlled vacuum (Figure 6.7). According to specification JTG E20—2011 T0620, the time required for the leading edge of the asphalt to pass between the start and stop marks is recorded and then the absolute viscosity can be calculated.

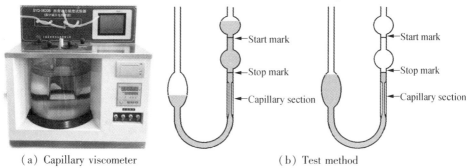

(a) Capillary viscometer　　　　　　　(b) Test method

Figure 6.7　Capillary viscosity test

(2) The rotational viscometer (RV) measures the torque required to rotate a spindle immersed in the sample fluid (Figure 6.8). It is mainly used to test the workability of asphalt at 135 ℃. According to specification JTG E20—2011 T0625, the test uses a rotational or Brookfield viscometer with a spindle placed in the asphalt sample and rotated at a specified speed. The material between the inner cylinder and the outer cylinder is analogous to the thin asphalt film in the sliding plate viscometer. The viscosity is determined by the amount of torque required to rotate the spindle at the specified speed.

Civil Engineering Materials

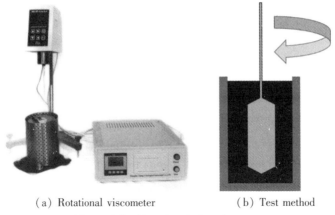

(a) Rotational viscometer   (b) Test method

Figure 6.8 Rotational viscosity test

(2) The sliding plate or microfilm viscometer measures the ratio of shear stress over the shear strain in a unit of time (Figure 6.9). It is mainly used to evaluate the viscosity of asphalt at low temperatures ranging from 10 to 55 ℃.

Figure 6.9 Sliding plate viscosity test

3) Kinematic viscosity

Kinematic viscosity is the ratio of absolute or dynamic viscosity divided by the density of the liquid at the temperature of measurement. Dynamic viscosity is a measurement of how difficult it is for a fluid to flow. The dynamic viscosity can be regarded as the friction inside asphalt, and the kinematic viscosity is the friction per unit weight.

6.3.4 Softening Point

Evaluation of the viscosity of asphalt requires a wide range of operating conditions including temperature, loading rate, stress, and strain. To simplify the test, the mechanical behavior and rheological properties of asphalt have been described using empirical tests and equations. The penetration test and the softening point test are the two most widely used empirical tests. According to specification JTG E20—2011 T0606, the softening point refers to the temperature at which the asphalt sample, heated at a controlled rate in a liquid bath, is soft enough to allow a 3.5g ball to fall a distance of 25 mm. It uses a copper ring and the heating rate is 5 ℃/min (Figure 6.10). If the softening point is higher than 80 ℃, we use an oil bath instead of a water bath. The penetration at the softening point is around 800 (0.1mm) and the viscosity at softening point is around 1200 Pa·s. The softening point, in combination with the penetration, can be used to evaluate the temperature susceptibility of asphalt.

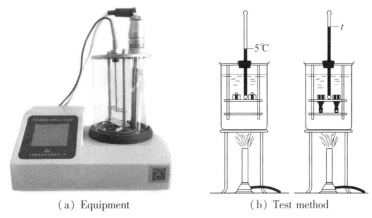

(a) Equipment   (b) Test method

Figure 6.10 Softening point test of asphalt

6.3.5 Brittle Point

According to specification JTG E20—2011 T0613, the brittle point is tested by brushing asphalt film on a piece of metal sheet and decreasing temperature (Figure 6.11). It refers to the temperature when asphalt film breaks. The penetration at the brittle point is around 1.2 (0.1 mm), therefore it is also called equivalent brittle point.

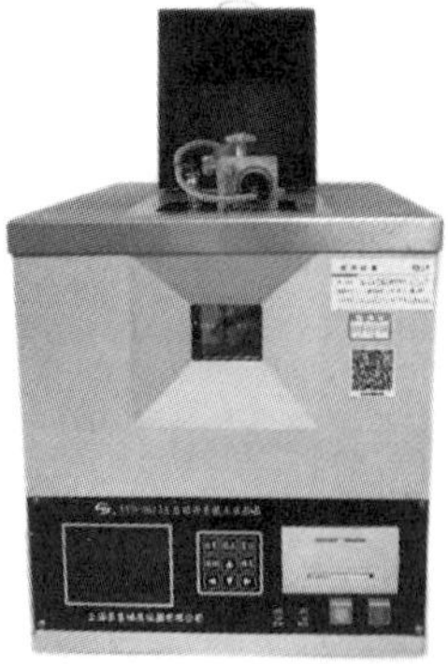

6.3.6 Ductility

According to specification JTG E20—2011 T0605, the ductility of asphalt refers to the stretched distance (cm) at breaking.

Figure 6.11 Asphalt brittle point tester

Ductility is tested by stretching a standard-sized dog-bone shape asphalt sample at the speed of 50 mm/min at 25 ℃ to its breaking point (Figure 6.12). Ductility at low temperatures such as 5 ℃ or 10 ℃ is an indicator of the resistance of the asphalt to cracking. Therefore, the ductility is also related to the viscosity of asphalt.

183

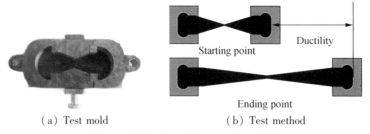

Figure 6.12 Ductility test of asphalt

6.3.7 Adhesion

According to specification JTG E20—2011 T0616, two tests can be used to evaluate the adhesion between asphalt and aggregates. The boiling method involves selecting five cubic shape particles with a size of 13.2 – 19 mm, coating the aggregate with asphalt, and boiling them in distilled water for 3 minutes. Adhesion is evaluated by 5 grades according to the situation of asphalt film peeling. The water immersion method involves mixing 100 g aggregates with a size of 9.5 – 13.2 mm and 5.5 g asphalt at a specified temperature to prepare the sample. After cooling, the mixture is immersed in distilled water at 80 ℃ for 30 minutes, and then the adhesion is evaluated according to the percentage of peeling area.

The adhesion of asphalt to an aggregate is dependent on complex factors. Carbonate aggregates such as limestone have a better affinity with asphalt than granite. The main characteristics of asphalt affecting its adhesion to aggregates are its viscosity, surface tension, and polarity. The viscosity and surface tension will govern the extent to which asphalt is absorbed into the pores at the surface of the aggregate particles. The absorption of asphalt into the aggregates also depends on the total volume of permeable pore space and the size of the pore openings.

6.3.8 Durability

Durability includes several different phenomena such as aging, moisture damage, frost damage, etc.

1) Aging

Aging refers to the change of properties and chemical composition during construction and the service life period of asphalt. It can be classified into short-term aging which occurs during the mixing, transportation, and paving process, and long-term aging which occurs in the field as a result of exposure to traffic and climatic conditions during its service life. Typical aging distress is cracking. Factors influencing aging include temperature, aggregate absorption of asphalt, and the presence of oxygen. Aging occurs most severely in hot climates, near pavement surfaces, in mixtures with more voids and high-absorptive aggregates.

2) Moisture damage

Moisture damage is mainly due to the stripping of asphalt and aggregates when encountering water. It is essential to dry aggregates thoroughly during asphalt mixture production to ensure good asphalt adhesion, and sometimes to add additives to facilitate the development of adhesion. The asphalt-aggregate interface is vulnerable to damage. The high density and low permeability of the mixture can improve moisture damage. Moisture damage will cause the rutting potential and material loss distress such as raveling and potholes on the asphalt pavement.

3) Frost damage

Frost damage is that the water in the voids of asphalt mixtures expands to form ice, causing pressures within the matrix of an asphalt mixture. In addition, the water in the voids during freezing and thawing also causes unusual dynamic pressures under traffic loading. Those pressures cause a local fracture at aggregate particle contacts and the material loss distress in the asphalt pavement. Practices suggest that a dense asphalt mixture with low content of air voids has a sufficiently low permeability to avoid severe moisture damage or frost damage.

6.3.9 Safety

At high temperatures, asphalt can flash or ignite in the presence of an open flame or spark. The flash point test is a safety test that measures the temperature at which asphalt flashes. The flash point is the lowest liquid temperature at which the application of the test flame causes the vapors of the sample to ignite. The Cleveland open cup method is usually adopted. According to specification JTG E20—2011 T0611, it is to partially fill the standard brass cup with an asphalt binder. The asphalt is then heated at a specified rate and a small flame is periodically passed over the surface of the cup, as shown in Figure 6.13. The flash point of asphalt

Figure 6.13 Cleveland open cup method

is the temperature that can sustain a flame for a short time after the volatile fumes come off the sample. The minimum temperature at which the volatile fumes are sufficient to sustain a flame for an extended time is the fire point. The flash point of an asphalt binder should be higher than 230 ℃.

6.4 Temperature Susceptibility

6.4.1 Temperature Dependency

The viscosity or consistency of asphalt is greatly affected by temperature. The asphalt gets hard and brittle at low temperatures and soft at high temperatures. The viscosity of asphalt decreases with the increase of temperature. There is an optimum range of viscosity for asphalt in the field at the annual temperature range. If the viscosity of asphalt is too high, the mixture will be too brittle and susceptible to low-temperature cracking. If the viscosity is too low, the mixture will flow readily, resulting in permanent deformation. There are also optimum viscosity ranges for mixing and paving. During construction, the asphalt needs to be heated to specific temperature ranges to meet the viscosity requirements for mixing and paving. As shown in Figure 6.14, the proper viscosity for mixing and paving mixtures are 0.17 and 0.25 Pa · s, respectively. The proper mixing temperature is 155 to 160 ℃ and the proper compaction temperature is 145 to 150 ℃. Warm mix additives can reduce the viscosity of asphalt and therefore the mixing and compaction can be done at a temperature lower than traditional hot mix asphalt mixture.

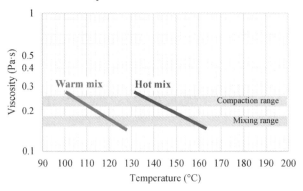

Figure 6.14　Viscosity for asphalt compaction and mixing

6.4.2 Time-temperature Equivalency

Asphalt is the typical time-temperature equivalent material. The time-temperature equivalent principle refers to the phenomenon that the time-dependent mechanical properties of materials rely on variations of temperatures. For example, the deformation in 1 hour at 60 ℃ is equivalent to the deformation in 10 hours at 25 ℃. Based on the time-temperature equivalency principle, the test results obtained at a higher temperature and shorter load duration are equivalent to tests performed at a lower temperature and longer load duration. Therefore, it is reasonable to predict the material's long-time mechanical properties based on their relationships with rising temperatures. It is also reasonable to shift creep curves at different temperatures into a master curve at a reference temperature.

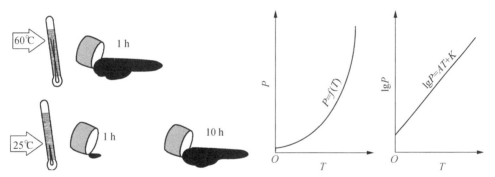

Figure 6.15　Time-temperature equivalency

Figure 6.16　Asphalt penetration vs. temperature

6.4.3 Penetration Index

There is a simple way to evaluate the temperature susceptibility of asphalt using the penetrations at different temperatures. The softening points are different temperatures measured at the same force and same deformation, while penetrations are the different deformations measured at the same force and same temperature. The results of the two tests are related. The penetration at the softening point is around 800 (0.1 mm). As shown in Figure 6.16, penetration has a near exponential relationship with temperature and the logarithm of penetration ($\lg P$) has a linear relationship with temperature (T).

$$\lg P = AT + K \tag{6.2}$$

where A = slope;
　　　K = intercept (constant).

The slope A represents the rate of change of penetration ($\lg P$) with temperature (T), indicating the temperature susceptibility. Therefore, A is called penetration-temperature susceptibility coefficient and can be calculated as Equation (6.3).

$$A = \frac{\lg 800 - \lg P_{(25\,°C,100\,g,5\,s)}}{T_{R\&B} - 25} \tag{6.3}$$

where $P_{(25\,°C,100\,g,5\,s)}$ = the penetration at 25 °C (0.1 mm);
$T_{R\&B}$ = softening point (°C).

For a very high-temperature susceptibility, A is close to infinity. For a very low-temperature susceptibility, A is close to zero. Based on Equation (6.4), we can also obtain the equivalent softening point temperature T_{800} at $P = 800$, and the equivalent brittle point temperature $T_{1.2}$ at $P = 1.2$, as shown in Equation (6.5).

$$T_{800} = \frac{\lg 800 - K}{A} = \frac{2.9031 - K}{A} \tag{6.4}$$

$$T_{1.2} = \frac{\lg 1.2 - K}{A} = \frac{0.0792 - K}{A} \tag{6.5}$$

The penetration index (PI) is defined as a function of the slope A, calculated as Equation (6.6). PI of -10 means a very high-temperature susceptibility, while PI of 20 means a very low-temperature susceptibility. PI can be used not only to evaluate the temperature sensitivity of asphalt, but also to classify the colloidal structure of asphalt. PI lower than -2 indicates a sol-type colloidal structure. PI between -2 and 2 indicates a sol-gel-type colloidal structure. PI higher than 2 indicates a gel-type colloidal structure.

$$PI = \frac{30}{1 + 50A} - 10 \tag{6.6}$$

PI is a very important property of asphalt. When designing asphalt mixtures for a climatic region, we should not only select asphalt of proper penetration, but also to check the penetration index. As shown in Figure 6.17, for a cold region, we want a higher penetration value, and asphalt A and B meet this requirement. However, asphalt A has a higher temperature susceptibility or a lower penetration index. Therefore, asphalt B with a lower temperature susceptibility or a higher penetration index is preferred.

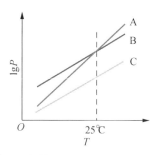

Figure 6.17 Asphalt with different penetration-temperature relationship

6.5 Aging and Modification

6.5.1 Aging

Aging is the change of properties and chemical composition of asphalt during construction and its service life period. There are two mechanisms of aging. Short-term aging occurs during mixing, transportation, placing, and compacting while the asphalt is at an elevated temperature, and is responsible for a significant change in properties. Long-term aging occurs gradually, and thus asphalt becomes ever harder during its service life. In general, as an asphalt binder ages, its viscosity increases and it becomes stiffer and more brittle, and causes cracking. Aging is a process including complicated physical changes and chemical reactions.

(1) Oxidation is the reaction of oxygen with asphalt. Many of the polar molecules within the asphalt are readily able to combine with any free oxygen they find. The result is additional cross-linking between molecules and a general stiffening and increase of viscosity.

(2) Volatilization is the evaporation of the lighter constituents of asphalt. It is primarily a function of temperature and occurs principally during hot mixed asphalt production.

(3) Polymerization is the combining of likely molecules to form larger molecules, causing a progressive hardening.

(4) Separation is the absorption of lighter saturate fractions by porous aggregates.

(5) Thixotropy is the setting of the asphalt when unagitated. Thixotropic effects can be somewhat reversed by heat and agitation. Asphalt pavements with little or no traffic are generally associated with thixotropic hardening.

(6) Syneresis is the separation of less viscous liquid from the more viscous asphalt molecular network. The liquid loss hardens asphalt and is caused by shrinkage or rearrangement of the asphalt structure due to physical or chemical changes. Syneresis is a form of bleeding.

As shown in Figure 6.18, short-term aging can be simulated by the thin-film oven (TFO), and rolling thin-film oven (RTFO). TFO simulates short-term aging by heating a film of asphalt for 5 hours at 163 ℃ (JTG E20—2011 T0609). For the RTFO aging, the asphalt binder is poured into the glass jars placed on a rack in a forced-draft oven at a temperature of 163 ℃ for 75 minutes (JTG E20—2011 T0610). The rack rotates vertically, continuously exposing the fresh asphalt. The asphalt in the rotating bottles is also subjected to an air jet to speed up the aging process. Long-term aging can be simulated by the pressure-aging vessel (PAV). It consists of a temperature and pressure-control chamber. A specified thickness of residue from the RTFO is placed in the PAV pans and then aged at the specified aging temperature for 20 hours under 2.1 MPa air pressure (JTG E20—2011 T0630). Since the pressure aging procedure forces oxygen into the sample, it is necessary to use a vacuum oven to remove any air bubbles from the sample before testing.

(a) TFO   (b) RTFO   (c) PAV

Figure 6.18　Asphalt aging equipment

The effects of aging can be evaluated by the change of composition and properties. With the increase of aging time, the amount of asphaltenes increases, and the amount of aromatics significantly decreases. A portion of the resins changes into asphaltenes while part of the aromatics changes into the resins. Because of the change of composition, the asphalt becomes stiffer. In terms of the change of asphalt properties, large differences in properties before and after aging indicate severe aging effects. The aging index refers to the ratio of asphalt properties after aging over those before aging. As shown in Figure 6.19, short-term aging is the aging that occurs during mixing, transportation, and

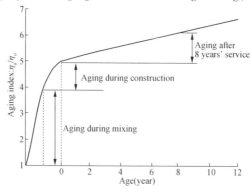

Figure 6.19　Asphalt aging process

application. It accounts for the majority of aging. Therefore, the warm-mix technology reduces the mixing temperature and can greatly reduce short-term aging.

6.5.2 Modification

Unmodified asphalt may be too soft at high temperatures, fracture at low temperatures, and lose adhesion under a combination of aging and moisture damage. Modification of asphalt is to add modifiers or additives into asphalt to improve the properties of asphalt or to add special properties to asphalt mixtures. Asphalt modification has been practiced for over 50 years and has achieved success. Table 6.3 summarizes the major types of modifiers. Some modifiers such as fillers, fibers, and antistripping additives can be directly added during mixing. However, many polymer modifiers need to be evenly mixed with asphalt to produce modified asphalt first. A shear mill can grind asphalt and modifier down to particles of between 1 μm and 3 μm during blending.

Table 6.3 Major types of asphalt modifiers

Types	Purposes	Examples
Filler	Reduce binder content, increase stability and bond	Mineral, fines, lime, cement, fly ash, carbon black
Extender	Reduce binder content	Sulfur, lignin
Rubber, and plastic	Increase stiffness at high temperature, increase elasticity at medium temperature, reduce stiffness at low temperature	Latex, SBS, SBR, crumb rubber, polyethylene
Fiber	Improve tensile strength	Asbestos, rock wool, mineral, fiber glass, polyester, cellulose
Antioxidant	Increase durability	Carbon, calcium salts
Hydrocarbon	Restore aged asphalt, increase stiffness	Recycled oil, hard/natural asphalt
Anti-stripping	Minimize stripping	Amines, lime

Styrene-butadiene-styrene (SBS), styrene-butadiene rubber (SBR), and ethyl vinyl acetate (EVA) are three of the most commonly used polymer modifiers. Polymer modification improves both the coating of the aggregates for improved durability and the elasticity of the binder, which benefits the rutting, fatigue, and thermal cracking resistance of the binder. Polymer blended with asphalt forms two phases, which are the polymer-rich phase and the asphalt-rich phase, depending on the content of the polymer. The key of rubber modified asphalt is the rubber-asphalt interaction including particle swelling and dissolution (Figure 6.20). Particle swelling is the light fractions such as aromatic oil of the asphalt are absorbed into the polymer chains of crumb rubber, forming a gel-like material that is double to triple the original volume. Dissolution is the chemical breakdown of polymer chains of rubber into small molecules and dissolving into the liquid phase of asphalt.

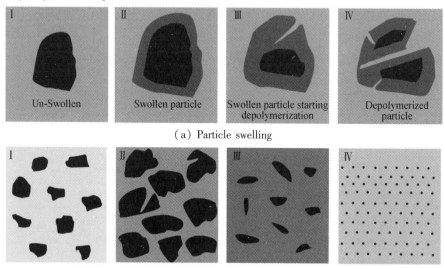

Figure 6.20 Mechanism of asphalt polymer modification

6.6 Penetration and Viscosity Grading

The asphalt binder is produced in several grades or classes. There are several grading systems for classifying asphalt binders, including penetration grading, viscosity grading, and performance grading.

6.6.1 Penetration Grading

The penetration grading system was developed in the early 1900s to characterize the consistency of asphalt. It is the most widely used asphalt grading system. It classifies asphalt mainly based on its penetration test results. The specification JTG F40—2004 defines seven grades of asphalt based on the penetration test results at 25 ℃ (Table 6.4).

6 Asphalt

Table 6.4 Penetration grades in China

Test items	Unit	Classes	160	130	110	90	70	50	30	Test methods
Penetration (25 °C, 5s, 100 g)	0.1 mm		140–200	120–140	100–120	80–100	60–80	40–60	20–40	T 0604
Climate region					2-1	2-2 3-2 1-1 1-2 1-3	2-2 2-3 1-3 1-4	2-2 2-3 2-4 1-4		
Penetration Index (PI)		A	−1.5 – +1.0							T 0604
		B	−1.8 – +1.0							
Min. softening point (R&B)	°C	A	38	40	43	45	46	49	55	T 0606
		B	36	39	42	43	44	46	53	
		C	35	37	41	42	43	45	50	
Min. dynamic viscosity at 60 °C	Pa·s	A	—	60	120	160	180	200	260	T 0620
Min. ductility at 10 °C	cm	A	50	50	40	45 30	20 20	15	10	T 0605
		B	30	30	30	20	15 15	10	8	
Min. ductility at 15 °C	cm	A, B	100	80	50	—	—	—	—	
		C	80	80	60	40 30	20	—	—	
Min. wax content	%	A	2.2							T 0615
		B	3.0							
		C	4.5							

Continued

Test items	Unit	Classes	Grades							Test methods
			160	130	110	90	70	50	30	
Min. flash point	°C			230		245		260		T 0611
Min. solubility	%			99.5		T 0607		—		—
Density	g/cm³			—		T 0603		—		—
			Residue of TFO (or RTFO) test							
Max. weight change	%		±0.8							T 0610 or T 0609
Max. penetration ratio	%	A	48	54	55	57	61	63	65	T 0604
		B	45	50	52	54	58	60	62	
		C	40	45	48	50	54	58	60	
Max. ductility at 10 °C	cm	A	12	12	10	8	6	4	—	T 0605
		B	10	10	8	6	4	2	—	
Max. ductility at 15 °C	cm	C	40	35	30	20	15	10	—	T 0605

The grades range from 30 to 160. Among them, grades 50 to 130 are the most widely used, while grade 30 is for the extremely hot area and grade 160 is for the extremely cold area. In each grade, there are three quality classes of asphalt from Class A to Class C. Class A is the best quality asphalt and can be used in any pavements and any layers. Class B can be used in the bottom layer of expressways and Class I roads, or all layers of lower-class roads. Class B can also be used as the base asphalt for modification, emulsification, and producing cutback asphalt. Class C can only be used in all layers of the Class III and lower-class roads.

In addition to testing the penetration of the original sample at 25 ℃, the specification JTG F40—2004 requires testing several other properties of asphalt. The penetration is to test the medium temperature performance. The penetration index is to test the temperature susceptibility. The softening point and kinematic viscosity at 60 ℃ are to test the high-temperature stability. Ductility at 15 ℃ and 10 ℃ is for low temperature cracking resistance. Wax content and solubility are for asphalt quality. The flash point is for safety reasons. Further, the weight change, penetration, and ductility of the short-term aged residues should also be tested to examine if the asphalt still has sufficient medium temperature penetration and low temperature cracking resistance after mixing and paving.

Climatic condition is a very important factor in selecting asphalt grade. According to specification JTG F40—2004, climatic regions are classified based on three criteria: the highest average temperature of one month, the lowest temperature, and the annual precipitation (Table 6.5). Based on the highest average temperature of one month, there are three climatic regions, i.e. very hot, hot, and cool. Based on the lowest temperature, there are four climatic regions, i.e. very cold, cold, cool, and warm. Based on the annual precipitation, there are four climatic regions, i.e. very humid, humid, dry, and very dry.

The climatic regions for selecting asphalt grade are named by two numbers. The first number is the high-temperature classification and the second number is the low-temperature classification. In addition to the temperature regions, rainfall regions are also defined. For example, the highest average temperature of one month in Jiangsu province is higher than 30 ℃, the lowest temperature is between −21.5 ℃ and −9 ℃, and the annual precipitation is 500 − 1000 mm. According to Table 6.5, Jiangsu province is at "1 − 3" temperature region and "2" rainfall region, meaning it is very hot in summer and cool in winter, and it is a humid area.

Table 6.5 Climatic regions for asphalt applications

	Regions	1	2	3	—
Hot climate region	High temperature	Very hot	Hot	Cool	—
	Highest average temperature of one month (℃)	>30	20–30	<20	
Cold climate region	Regions	1	2	3	4
	Low temperature	Very cold	Cold	Cool	Warm
	Lowest temperature (℃)	<−37	−37 – −21.5	−21.5 – −9	>−9
Rainfall region	Regions	1	2	3	4
	Rainfall	Very humid	Humid	Dry	Very dry
	Annual precipitation (mm)	>1000	500-1000	250-500	<250

6.6.2 Viscosity Grading

In the early 1960s, an improved asphalt grading system was developed based on a rational scientific viscosity test. This scientific test replaced the empirical penetration test as the key asphalt characterization and is now still used in areas like Canada. Viscosity grading can be done on original asphalt samples, called AC grading, or aged residue samples, called AR grading. The AR grading system is an attempt to simulate asphalt properties after short-term aging and thus, it should be more representative of how asphalt behaves in HMA pavements.

In the viscosity grading system, the 60 ℃ viscosity is to test the asphalt performance at high pavement temperature. The 135 ℃ viscosity is to test the asphalt performance at mixing temperature. Penetration and ductility at 25 ℃ are to test the asphalt performance at medium pavement temperature. The flash point is to test asphalt safety. Solubility is to test asphalt quality. Besides, the viscosity and ductility test the performance of short-term aged asphalt at high and medium pavement temperature, respectively.

Details of the AC and AR viscosity grading can be found in the specification AASHTO M226-80 and the ASTM D3381/D3381M-13. As shown in Table 6.6, for AC viscosity grading, AC-5 grade means the viscosity at 60 ℃ is 500 ± 100 poise, which is less viscous than the AC-40 grade whose 60 ℃ viscosity is 4000 ± 800 poise. Typical grades used for hot mixed asphalt paving in the US are AC-10, AC-20, AC-30, AR-4000, and AR-8000.

Table 6.6 Viscosity grading

Specifications	Grading based on original asphalt (AC)						Grading based on aged residue (AR)				
AASHTO M226	AC-2.5	AC-5	AC-10	AC-20	AC-30	AC-40	AR-10	AR-20	AR-40	AR-80	AR-160
ASTM D3381	AC-2.5	AC-5	AC-10	AC-20	AC-30	AC-40	AR-1000	AR-2000	AR-4000	AR-8000	AR-16000

Figure 6.21 shows the relationships between penetration grading, AC viscosity grading, and AR viscosity grading. Asphalt with the same viscosity might be classified into three penetration grades. Therefore, the empirical penetration grading sometimes may not effectively differentiate asphalt. The viscosity grading is more rational.

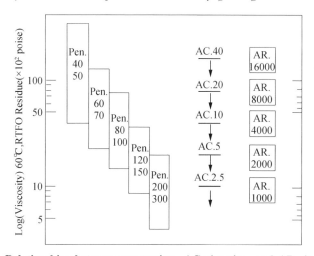

Figure 6.21 Relationships between penetration, AC viscosity, and AR viscosity grading

6.7 Performance Grading

6.7.1 Equipment

In 1987, the Strategic Highway Research Program (SHRP) began developing a new system for specifying asphalt materials and designing asphalt mixes. The SHRP produced the Superpave, which is the Superior Performing Asphalt Pavements mix design method for asphalt concrete and the performance grading method for asphalt specification (AASHTO M 320 − 10). The performance grade, or PG grade, classifies asphalt by performance at the applicable temperature of the region where the road is located. It involves a grading discipline and a series of new testing methods.

1) RTFO

The RTFO is used to simulate short-term aging (Figure 6.18(b)). It is not newly developed equipment. In the test, asphalt is poured into the glass jars placed on a rack in a forced-draft oven at a temperature of 163 ℃ for 75 min. The rack rotates vertically. The asphalt in the rotating jars is subjected to an air jet to speed up the aging process.

2) PAV

The PAV can simulate 5 to 10 years of long-term aging (Figure 6.18(c)). It consists of a temperature and pressure-control chamber. A specified thickness of residue from the RTFO is placed in the PAV pans and then aged at the specified aging temperature for 20 hours under 2.1 MPa. The aging temperature ranges from 90 to 110 ℃ and is selected according to the grade of the asphalt.

3) RV

The RV tests the workability for mixing and paving of original asphalt at 135 ℃ (Figure 6.8). The viscosity is determined by the amount of torque required to rotate the spindle at the specified speed. The spindle size used is determined based on the viscosity being measured.

4) Dynamic shear rheometer (DSR)

The DSR is used in all stages of the PG tests to evaluate both rutting and fatigue potentials of asphalt (Figure 6.22). According to specification JTG E20—2011 T0628, one of the parallel plates oscillates with respect to the other at preselected frequencies

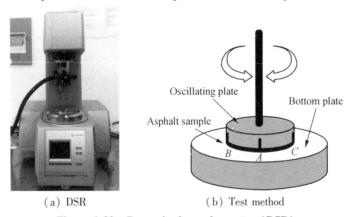

(a) DSR (b) Test method

Figure 6.22 Dynamic shear rheometer (DSR)

and rotational deformation amplitudes or torque amplitudes. For evaluating rutting potential, the test temperature is equal to the high temperature for the grade of asphalt and the sample size is 25 mm in diameter and 1 mm in thickness. For evaluating fatigue potential, the intermediate temperature is used and the sample size is 8 mm in diameter and 2 mm in thickness.

Complex shear modulus (G^*) and phase angle (δ) can be obtained from the DSR test. Based on the shear stress-time and the shear strain-time curves, G^* is the ratio of the maximum shear stress (τ_{max}) over the maximum shear strain (γ_{max}). The complex shear modulus (G^*) can be considered as the sample's total resistance to deformation when repeatedly sheared. The phase angle (δ), is the lag between the applied shear stress and the resulting shear strain (Figure 6.23). The larger the phase angle, the more viscous the material. For pure elastic material, phase angle equals 0.

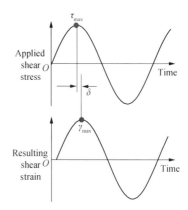

Figure 6.23 Shear stress and strain during the DSR test

$$G^* = \frac{\tau_{max}}{\gamma_{max}} \tag{6.7}$$

where τ_{max} = the maximum shear stress (MPa);

γ_{max} = the maximum shear strain.

Because of the viscoelastic nature of asphalt, G^* is composed of two parts: the elastic or storage modulus (G') and the viscous or lost modulus (G'') (Figure 6.24).

$$G' = G^* \cos\delta \tag{6.8}$$
$$G'' = G^* \sin\delta \tag{6.9}$$

where G' = elastic (storage) modulus (MPa);

G'' = viscous (lost) modulus (MPa).

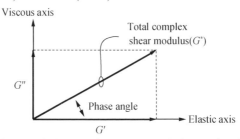

Figure 6.24 Elastic (storage) modulus and viscous (lost) modulus

G^* and δ are used as predictors of HMA rutting and fatigue cracking. Rutting is the main concern in the early service life of pavement, while fatigue cracking becomes the major concern in the late service life. Table 6.7 presents the PG requirements of the DSR test results, which are mainly to address two kinds of distress.

(1) To address rutting, $G^*/\sin\delta$ should be no less than 1 kPa and 2.2 kPa for the original binder and RTFO residue, respectively. To resist rutting, the asphalt should be stiffer and more elastic. Intuitively, high G^* means the asphalt is stiffer. Low δ means the asphalt is elastic and able to recover its original shape after being deformed by a load. Therefore, a minimum value of $G^*/\sin\delta$ is specified.

(2) To address fatigue cracking, $G^*\sin\delta$ should be no less than 5000 kPa. To resist fatigue cracking, the asphalt should be elastic and able to dissipate energy by rebounding and not cracking. But it should not be too elastic or too stiff which may cause cracking rather than deform-then-rebound. As shown in Figure 6.24, $G^*/\sin\delta$ is the viscous modulus G''. A smaller viscous modulus means less deformation or damage at each load and therefore is good for preventing fatigue cracking. Therefore, when fatigue cracking is of greatest concern late in a pavement's life, a maximum value for the viscous component of the complex shear modulus is specified.

Table 6.7 PG criteria of the DSR test results

Materials	Value	Criteria	Distress
Original binder	$G^*/\sin\delta$	⩾1.0 kPa	Rutting
RTFO residue	$G^*/\sin\delta$	⩾2.2 kPa	Rutting
PAV residue	$G^*\sin\delta$	⩽ 5000 kPa	Fatigue cracking

5) Bending beam rheometer (BBR)

The BBR is used to test the thermal cracking potential of short-term and long-term aged asphalt. According to specification JTG E20—2011 T0627, it measures the midpoint deflection of a simply supported prismatic beam of asphalt subjected to a constant load applied to its midpoint (Figure 6.25). The test temperature equals the low-temperature rating plus 10 ℃. The flexural creep stiffness of the beam is calculated based on the load magnitude, deflection, and dimensions of the beam specimen.

$$S(t) = \frac{PL^3}{4bh^3\Delta(t)} \tag{6.10}$$

where P = load, 100 g;

L = support span, 102 mm;
b = beam width, 12.7 mm;
h = beam thickness, 6.35 mm;
$\Delta(t)$ = deflection at 60 s (mm).

(a) BBR

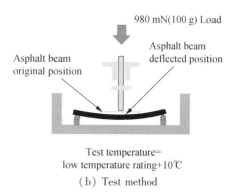
(b) Test method

Figure 6.25　The BBR test

As the beam creeps, the midpoint deflection is monitored. The flexural creep stiffness of the beam is calculated by dividing the maximum stress by the maximum strain for each of the specified loading times. The low-temperature thermal-cracking performance of asphalt mixtures is related to the creep stiffness and the slope of the logarithm of the creep stiffness versus the logarithm of the time curve of the asphalt (Figure 6.26).

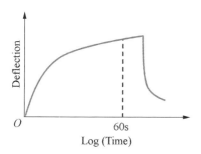
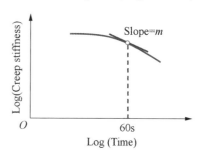

Figure 6.26　The creep stiffness and slope

6) Direct tension tester (DDT)

The DDT measures the failure strain and failure stress of a dog-bone shape asphalt specimen under direct tension (Figure 6.27). According to specification JTG E20—2011 T0629, The test temperature is between -36 ℃ and 0 ℃. The strain at failure is a measure of the amount of elongation that the asphalt can sustain without cracking. It is used as a criterion for specifying the low-temperature properties of the binder.

$$\varepsilon_f = \frac{\Delta L}{L_e} \tag{6.11}$$

where ε_f = failure strain.

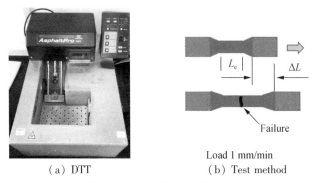

(a) DTT   (b) Test method

Figure 6.27   The DDT test

6.7.2 Testing Temperature

Unlike previous specifications that require performing the test at a fixed temperature and varying the requirements for different grades of asphalt, the PG specifications require performing the test at different critical pavement temperatures and fixing the criteria for all grades. The binder performance grading and Superpave mix design methods include performance-based specifications for asphalt binders and mixtures to primarily control three kinds of distress: rutting, thermal cracking, and fatigue. To address the three main kinds of distress, three pavement design temperatures are required for the PG grade: a maximum, a minimum, and an intermediate temperature (Table 6.8). The maximum pavement design temperature is the highest successive 7-day average maximum pavement temperature. The minimum pavement design temperature is the minimum

Table 6.8   PG testing temperatures

Distress	Temperature levels	Pavement design temperatures	Determinations	Test temperatures
Rutting	Maximum	Highest successive 7-day average maximum pavement temperature	20 mm below the pavement surface	Same
Thermal cracking	Minimum	Minimum surface temperature expected over the life of a pavement	Pavement surface	+10 ℃
Fatigue	Intermediate	Average of maximum and minimum pavement design temperature +4 ℃	Average temperature +4 ℃	Same

surface temperature expected over the life of a pavement. The intermediate pavement design temperature is the average of maximum and minimum pavement design temperature + 4 ℃. The maximum and intermediate test temperatures are the same as their pavement design temperatures. The minimum test temperature is the minimum pavement design temperature +10 ℃ to reduce the testing time.

6.7.3 PG Grades

Names of grades start with PG, which means performance graded, followed by two numbers representing the maximum and minimum pavement design temperatures in Celsius. Table 6.9 shows the binder grades in the PG specifications. The high-temperature grade ranges from 46 ℃ to 82 ℃ while the low-temperature grades range from −46 ℃ to −10 ℃. It is noted that PG asphalt binders are specified in increments of 6 ℃.

Table 6.9 Binder grades in the PG specifications

High temperature grades (℃)	Low temperature grades (℃)
PG 46	−34, −40, −46
PG 52	−10, −16, −22, −28, −34, −40, −46
PG 58	−16, −22, −28, −34, −40
PG 64	−10, −16, −22, −28, −34, −40
PG 70	−10, −16, −22, −28, −34, −40
PG 76	−10, −16, −22, −28, −34
PG 82	−10, −16, −22, −28, −34

PG binders that differ in the high- and low-temperature specifications by 90 ℃ or more generally require some sort of modification. As shown in Figure 6.28, asphalt of PG grades on the left top can be directly obtained from a crude oil refinery. The asphalt

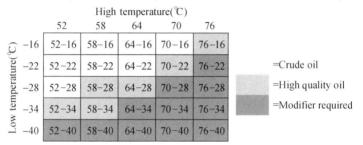

Figure 6.28 Asphalt of different PG grades

Civil Engineering Materials

of darker PG grades in the middle are from high-quality crude oil. Asphalt of PG grades on the bottom right is modified asphalt, which can be used in regions with very high temperature differences.

6.7.4 PG Tests

The main PG grading tests of asphalt can be classified into three stages (Table 6.10). The first stage is to test the workability of the original binder using RV. The second stage is to evaluate the rutting, fatigue, and thermal cracking risks of the short-term RTFO aged asphalt residue, using DSR, BBR, and DTT. The third stage is to evaluate the fatigue and thermal cracking performance of long-term PAV aged asphalt residue, using DSR, BBR, and DTT. It is noted that DSR can be used to evaluate both rutting and fatigue and is the most important equipment in the Superpave PG grading system.

Table 6.10 PG grading tests of asphalt

Tests	Original binder	Short-term aged (RTFO)	Long-term aged (PAV)
Rotational viscometer (RV)	Workability	—	—
Dynamic shear rheometer (DSR)	—	Rutting & fatigue	Fatigue
Bending beam rheometer (BBR)	—	Thermal cracking	Thermal cracking
Direct tension tester (DTT)	—	Thermal cracking	Thermal cracking

Figure 6.29 summarizes the PG grading tests of asphalt and addressed properties including workability, rutting, fatigue cracking, and thermal cracking potential. Firstly,

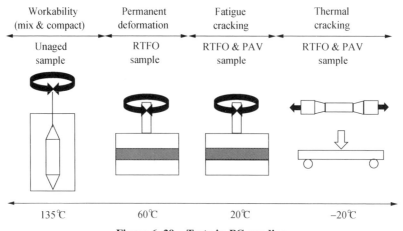

Figure 6.29 Tests in PG grading

the rotational viscosity at 135 ℃ of the unaged sample is tested to evaluate the workability of a sample. Secondly, the rutting potential of the RTFO aged sample is tested using DSR at the highest pavement design temperature. The fatigue cracking potential of both RTFO and PAV aged samples is tested using DSR at the intermediate pavement design temperature. The thermal cracking potential of both RTFO and PAV aged samples is tested using BBR and DTT at the minimum pavement design temperature plus 10 ℃.

Table 6.11 summarizes the specifications or test temperatures for the PG grading system. The physical properties or criteria remain constant for all grades, but the temperatures at which these properties must be achieved vary, depending on the climate at which the binder is expected to be used.

6.7.5 Selection of PG Grades

As shown in Figure 6.30, the proper asphalt grade can be selected in seven steps:

(1) Determining the distributions of the annual 7-day maximum air temperatures (50 ℃) and 1-day minimum air temperatures (Figure 6.30(a)).

(2) Determining pavement temperatures based on air temperatures (Figure 6.30(b)).

(3) Determining the average 7-day maximum (51 ℃) and the average 1-day minimum (−17 ℃) pavement temperatures (Figure 6.30(c)).

(4) There is a 50% probability that the maximum pavement temperature exceeds 51 ℃ or the minimum pavement temperature is lower than −15 ℃. Therefore, the PG grade of asphalt should be larger than this temperature range (Figure 6.30(d)).

(5) Since PG asphalt binders are specified in increments of 6 ℃, the nearest grade is PG 52−22. This covers a wider range of temperatures, and therefore its reliability will be more than 50% (Figure 6.30(e)).

(6) For a 98% reliability, the asphalt must cover the range from −23 to 55 ℃. This means that there is a 2% probability that the high and low pavement temperature exceeds 50 ℃ or be lower than −23 ℃ (Figure 6.30(f)).

(7) The closest PG asphalt binder grade is PG 58−28. Therefore, a PG 58−28 is selected to meet 98% reliability requirements. Since this covers a slightly wider range of temperatures, its reliability will be more than 98% (Figure 6.30(g)).

Table 6.11 Specifications and test temperatures for the PG grading system

Performance grades	PG 46-				PG 52-							PG 58-								PG 64-								PG 70-							PG 76-							PG 82-				
	34	40	46	10	16	22	28	34	40	46	16	22	28	34	40	46	16	22	28	34	40	10	16	22	28	34	40	10	16	22	28	34	10	16	22	28	34								
Average seven-day maximum pavement design temperature (℃)	<46				<52							<58								<64								<70							<76							<82				
Minimum pavement design temperature (℃)	-34	-40	-46	-10	-16	-22	-28	-34	-40	-46	-16	-22	-28	-34	-40	-46	-16	-22	-28	-34	-40	-10	-16	-22	-28	-34	-40	-10	-16	-22	-28	-34	-10	-16	-22	-28	-34								
	Original binder																																												
Min. flash point temperature (℃)	230																																												
Viscosity, ASTM D4402: Maximum, 3Pa·s, test temperature (℃)	135																																												
Min. $G^*/\sin\delta$, 1 kPa test temperature at 10 rad/s (℃)	46				52							58								64								70							76							82				
	RTFO residue																																												
Max. weight loss (%)	1																																												
Min. $G^*/\sin\delta$, 2.2 kPa Test temperature at 10 rad/s (℃)	46				52							58								64								70							76							82				
	Pressure-aging vessel residue																																												
PAV aging temperature (℃)	90				90							100								100								100 (110)							100 (110)							100 (110)				
Max. $G^*/\sin\delta$, 5000 kPa test temperature at 10 rad/s (℃)	10	7	4	25	22	19	16	13	10	7	25	22	19	16	13	10	31	28	25	22	19	34	31	28	25	22	19	37	34	31	28	25	40	37	34	31	28								
Max. creep stiffness: S, 300 MPa, Min. m-value, 0.3 test temperature at 60 s (℃)	-24	-30	-36	0	-6	-12	-18	-24	-30	-36	-6	-12	-18	-24	-30	-36	-6	-12	-18	-24	-30	0	-6	-12	-18	-24	-30	0	-6	-12	-18	-24	0	-6	-12	-18	-24								
Min. direct tension: failure strain, 1.0% test temperature at 1.0 mm/min (℃)	-24	-30	-36	0	-6	-12	-18	-24	-30	-36	-6	-12	-18	-24	-30	-36	-6	-12	-18	-24	-30	0	-6	-12	-18	-24	-30	0	-6	-12	-18	-24	0	-6	-12	-18	-24								

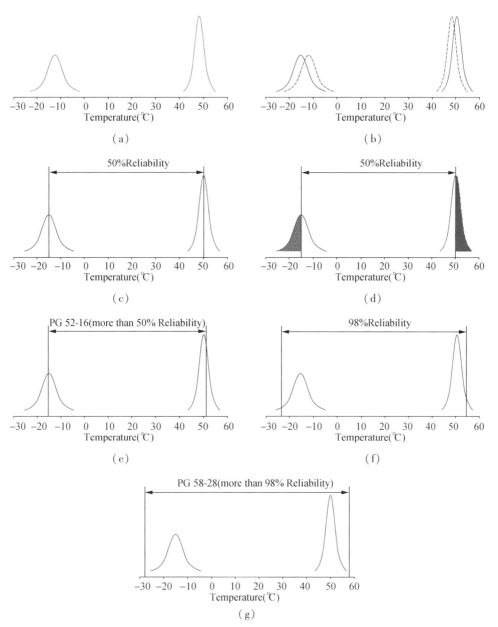

Figure 6.30 Selection of asphalt PG grades

Questions

1. Discuss three asphalt production methods and discuss if the produced asphalt can be directly used in the pavement?
2. What are the mechanisms of asphalt cutbacks and asphalt emulsions, respectively?
3. Discuss the chemical composition and colloidal structure of asphalt.
4. Briefly discuss the procedures of penetration test and softening point test.
5. Define what is the temperature susceptibility of asphalt and draw a graph showing viscosity versus temperature of asphalt?
6. Discuss how to obtain the penetration index of an asphalt sample based on the results of the penetration and softening point test.
7. Discuss the aging of asphalt during mixing and in service. How to simulate the different types of aging of asphalt in the laboratory?
8. What is the purpose of each of these tests?
   (1) Flash point test
   (2) RTFO procedure
   (3) RV test
   (4) DSR test
   (5) BBR test
9. Define the three methods used to grade asphalt. Which method is used in your country?
10. What kind of tests and samples (original, short-term aged, and long-term aged) are used in the PG grade to evaluate the workability, rutting, fatigue cracking, and thermal cracking, respectively?

7 Asphalt Mixtures

The asphalt mixture, also known as asphalt concrete, is a material mainly composed of asphalt and aggregates to achieve specific properties or performance useful for the factory process or the performance during the life cycle service. A mineral filler may be added to aggregates to improve the quality and performance of asphalt concrete, such as the resistance to fatigue damage, low-temperature cracking, and moisture damage. An asphalt mixture consists of 90 %–95 % aggregates, and its performance is greatly influenced by the characteristics of the aggregates. Additives and admixtures usually accounting for less than 3 % of the total mixture are added to asphalt mixtures to enhance performance or workability. Asphalt mixtures feature many properties making them ideal materials for pavement. Asphalt mixtures have sufficient high strength and surface texture for comfortable riding. It needs short time construction and traffic which can be immediately opened after construction. It is also very easy to maintain. The main drawbacks of asphalt mixtures are temperature and moisture susceptibility, and therefore a careful mixture design is needed. This chapter introduces the classification, structure, properties, volumetric parameters, mix design, and construction of asphalt mixtures.

7.1 Classification

7.1.1 Type of Mixture

Asphalt mixtures can be classified in many different ways. Asphalt materials in pavement engineering can be classified as asphalt mixtures which include mixing asphalt with aggregates, or asphalt seals which involve spraying asphalt binder with or without aggregate chips. Asphalt mixtures can be used on all classes of roads while asphalt seals are mainly for lower-class roads or as tack coats.

1) Asphalt mixtures

Asphalt mixtures refer to the mixtures made by drying and heating the aggregate of a designed gradation to a specified temperature and mixing the asphalt with certain viscosity in the specified proportion at the given temperature. Asphalt mixtures are suitable for all classes of pavements and are widely used in both road and airport construction. One important benefit of the asphalt mixtures is that its strength forms quickly after the surface layer is compacted and cooled and therefore the pavement can open the traffic within a few hours.

2) Asphalt seals

The asphalt seal or seal coat is to apply a protective coating made from emulsified asphalt with or without aggregate chips on aged asphalt pavement to restore its functional performance. The emulsified asphalt is soaked into the asphalt pavement and fills in small gaps in the aggregate for a smoother appearance. A fog seal is a light application of a diluted slow-setting asphalt emulsion to the surface of an aged or oxidized pavement surface. It takes a long time to open to the traffic after application. An excessive amount of asphalt will form a film on the pavement surface, causing low skid resistance and therefore the spraying rate must be strictly controlled. Fog seals are usually used on roads and parking lots with low traffic and speed. Sand seals can be used to improve skid resistance.

7.1.2 Temperature

Based on the mixing and paving temperature, asphalt mixtures can be classified into hot, warm, and cold mixtures. Hot mixed asphalt (HMA) is mixed and paved at 120 – 160 ℃. The asphalt binder used in the HMA is not fluid at room temperature, so once HMA cools after paving, it becomes solid and strong enough to support traffic loads.

A warm mixture can be mixed and paved at 80 – 120 ℃. It uses physical and chemical methods including emulsifying, foaming, and adding additives to reduce the viscosity of asphalt at a temperature lower than HMA. Warm mixtures have lots of benefits including less energy and fuel consumption to heat the mix, fewer greenhouse gases, and less oxidation of the asphalt binder. The lower mixing temperatures can be achieved by several techniques including (1) using chemical additives to lower the high-temperature viscosity of the asphalt binder, (2) adding water to the binder, causing it to become foamy asphalt or asphalt foam, and (3) Using a two-stage process which involves the addition of hard and soft asphalt at different points during mix production. The warm mixture has several benefits. The cost is lower because significantly less fuel is required to bring the mix to the desired temperature. Therefore, the emissions are also lower and there is a decreased environmental impact. In addition, there is the potential for improved performance because of the decreased time for hardening.

Cold mixtures can be used above 10 ℃. It uses cutback asphalt or emulsified asphalt. The solvent in the cutback or water in the emulsion evaporates, leaving the asphalt residue as the binder. Emulsified asphalt is widely used for asphalt sealing coats. Cutbacks have been widely used for pavement construction. However, it is mainly used for pothole patching now due to the cost of expensive solvents and the volatility of hazardous solvents.

7.1.3 Gradation

1) Dense gradation

The dense-graded mixture uses large, medium, and fine sizes of aggregates packing densely. The air void content of dense graded mixtures ranges from 3 % to 6 %. Dense-graded mixtures are the most common asphalt mixture type and can be used in any layer of the pavement structure for any traffic levels. The term dense-graded refers to the dense aggregate gradation, which means that there is relatively little space between the aggregate particles in such mixtures. The dense gradation is also called continuous dense gradation or the suspended-dense structure in which the coarse particles suspend in finer particles (Figure 7.1(a)). Therefore, its skeleton is not very strong and the air voids content is low. Properties of this structure include low permeability, high cohesion, low friction, high crack resistance, better durability, and poor high-temperature stability.

2) Gap gradation

The gap-graded mixture uses large and fine sizes of aggregates with few medium size aggregates. The air void content of gap-graded mixtures also ranges from 3 % to 6 %. The large aggregates in the gap-graded mixtures can form a stone-on-stone skeleton. The stone matrix asphalt (SMA) or stone mastic asphalt is a typical gap-graded mixture. The gap gradation is also called the skeleton-dense structure, in which coarse aggregates contact each other to form a strong skeleton (Figure 7.1(b)). It needs more asphalt to form asphalt mortar with fine aggregates to fill the voids left by coarse aggregates. This structure has high cohesion and skeleton friction. Therefore, it has good performance at both low and high temperatures. The durability is also good because of low air voids. However, because of the special gradation, it should be cautious for segregation during the paving process.

3) Open gradation

The open-graded mixture uses few fine aggregates, leaving many interconnected air voids in the mixture. The air void content of the open-graded mixture is larger than 18% and therefore it is used as the porous asphalt mixture. The open gradation is also called the continuous open gradation or the skeleton-open structure, in which coarse aggregates contact each other, but there are very few fine aggregates to fill the voids between coarse aggregates, leaving a lot of interconnected air voids (Figure 7.1(c)). Therefore, it has a very high permeability. This structure has high interconnected air voids, but it is moisture susceptible and has low fatigue resistance and durability. It needs very good quality asphalt with high adhesion and viscosity.

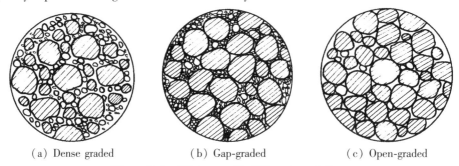

(a) Dense graded   (b) Gap-graded   (c) Open-graded

Figure 7.1　Mixtures with different gradations

7.1.4　Aggregate Size

The asphalt pavement construction specification JTG F40—2004 classifies asphalt mixtures into super coarse, coarse, medium, fine, and sand mixes based on the nominal maximum aggregate size (NMAS) of the aggregate (Table 7.1). The thresholds of aggregate size to classify those mixtures are 9.5, 16, 26.5, and 37.5 mm, respectively. Usually, half of the nominal maximum aggregate size is used as the threshold to differentiate coarse and fine aggregates in mixtures. Table 7.1 shows the classification and terminology of asphalt mixtures used in China. The different mixture types are employed to satisfy the different demands of pavement performance and also to accommodate the varying nature of the aggregates and asphalt supplies.

Table 7.1 Classification of asphalt mixtures in JTG F40—2004

Mixtures	Asphalt concrete	Asphalt treated base	Stone matrix asphalt	Asphalt macadam	Asphalt treated porous base	Open-graded friction course	NMAS (mm)	Maximum aggregate size (mm)
Gradation	Dense	Dense	Gap	Semi-open	Open	Open		
Permeability	Impermeable	Impermeable	Impermeable	Semi-permeable	Permeable	Permeable		
Binder content(%)	4–8	3–5	5–7	2–5	2–5	3–6		
Void(%)	3–5	3–6	3–4	6–12	≥18	≥18		
Super Coarse	—	ATB-40	—	—	ATPB-40	—	37.5	53
Coarse	—	ATB-30	—	—	ATPB-30	—	31.5	37.5
Coarse	AC-25	ATB-25	—	—	ATPB-25	—	26.5	31.5
Medium	AC-20	—	SMA-20	AM-20	—	—	19	26.5
Medium	AC-16	—	SMA-16	AM-16	—	OGFC-16	16	19
Fine	AC-13	—	SMA-13	AM-13	—	OGFC-13	13.2	16
Fine	AC-10	—	SMA-10	AM-10	—	OGFC-10	9.5	13.2
Sand	AC-5	—	—	AM-5	—	—	4.75	9.5

7.2 Composition and Strength

In pavement design, the asphalt mixture is usually treated as a homogenous material with specific mechanical properties. However, to better understand the mechanism of distress, it is important to examine and treat the asphalt mixture as a heterogeneous composite considering the spatial distribution of the different component materials and their void characters. An asphalt mixture is a composite of randomly oriented and distributed aggregate particles bonded together by a continuous matrix of asphalt. The two phases are quite different. While the aggregate is stiff and hard, asphalt is flexible and soft and is particularly susceptible to temperature change. Therefore, the proportion of asphalt in the mixture has a great influence on the mixture's properties. The unique and complex characteristics of the asphaltic phase appear to be of major importance because they introduce a rate and temperature dependence to the mechanical properties of the whole composite. The skeleton of mineral aggregates also greatly influences the properties of the asphaltic composite because of its high volume fraction in the mixtures and the elevated stiffness. The aggregate skeleton plays a very important role in transferring the traffic loads to lower layers of pavement. Further, the void content and distribution are also of great importance to the performance of the mixture.

7.2.1 Composition

There are two theories for the structure of asphalt mixtures. The surface theory regards the asphalt mixture as the aggregate skeleton and a binder which coats the surface of all aggregates. The skeleton is composed of coarse aggregates, fine aggregates, and mineral fillers. The mastic theory regards the asphalt mixture as a multiscale gel dispersed structure system. As shown in Figure 7.2, at the macro-scale, the asphalt mixture is a system in which coarse aggregates disperse in asphalt mortar. At the meso-scale, asphalt mortar is a system in which fine aggregates disperse in asphalt mastic. At the micro-scale, asphalt mastic is a system in which mineral fillers disperse in the asphalt binder.

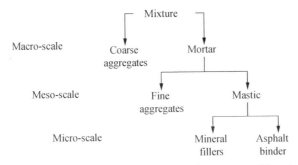

Figure 7.2 Structure of the asphalt mixture at multi-scales

Therefore, an asphalt mixture is a composite comprised of coarse aggregates distributed within asphalt mortar. The asphalt mortar is a homogenous mix of the asphalt binder, fine aggregates, and mineral fillers. Its properties inherit that of asphalt mastic and are also governed by the sand-mastic ratio. Increasing the content of coarse aggregates, the randomly oriented and distributed coarse aggregate particles will come close to the state of stone-on-stone contact, and the mixture will form a stable skeleton. Under given compaction effort, the mixture can be impermeable or permeable and even be open, depending on the ratio of asphalt mortar and voids in the compacted coarse aggregates.

The aggregate skeleton structure has a great impact on the properties of the asphalt mixture. As we have learned earlier, asphalt mixtures generally have three gradations: dense, gap, and open gradation. Previously in the United Kingdom and many other countries, asphalt mixtures used to be classified into two groups, namely, "asphalt concrete" and "asphalt coated macadam". Figure 7.3 illustrates the fundamentally different characteristics of asphalt concrete and asphalt macadam. Asphalt concrete relies on its dense, stiff mortar for strength and stiffness, whereas asphalt macadam relies on the stability of the aggregate through its grading. Thus, the role of asphalt is quite different in each case, and the properties of mixtures are more strongly dependent on the nature of the asphalt than the properties of macadam.

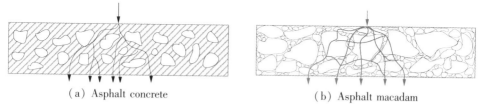

(a) Asphalt concrete   (b) Asphalt macadam

Figure 7.3   **Essential features of asphalt concrete and asphalt macadam**

7.2.2 Strength Parameters

Two different mechanisms of permanent deformation in the asphalt mixture are densification and plastic shear deformation. As the name suggests, densification occurs due to an overall change in the volume of the material due to the repeated action of the wheel loads. The contribution of densification to the overall permanent deformation of an asphalt mixture is typically small. The permanent deformation due to the accumulation of plastic shear strain has a much more significant contribution to rutting. To characterize the shear strength of the asphalt mixture, we can borrow the concept of cohesion c and internal friction angle φ in the soil mechanics. According to the Mohr-Coulomb failure criterion shown in Figure 7.4, the shear strength (τ) of an asphalt mixture at a point on a particular plane is expressed as a linear function of the normal stress (σ) on the plane at the same point.

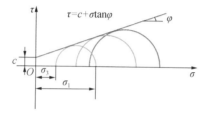

Figure 7.4   **A Mohr-Coulomb envelope diagram**

In the asphalt mixture, the asphalt binder is mainly responsible for cohesion while the aggregate skeleton is mainly responsible for the internal friction angle. In Figure 7.5, mixtures A and B have the same internal friction angle, but the cohesion of A is higher, indicating the two mixtures may use the same gradation but A has a better quality asphalt. Mixtures C and D have the same cohesion, but the internal friction angle of C is higher, indicating the two mixtures may use the same asphalt binder but the gradation of C provides a stronger skeleton.

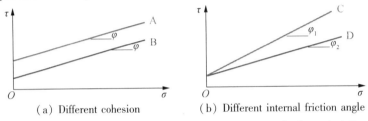

Figure 7.5 Asphalt mixtures of different shear strength characteristics

The determination of parameters c and φ can be obtained through the direct shear test of the asphalt mixture, which is very similar to the direct shear test of soils. By measuring the shear strength under different pressure, the Coulomb line can be drawn in the $\tau - \sigma$ chart to obtain the c and φ values. Triaxial test, and the simple tensile-compressive test can also be carried out to obtain the cohesion c and internal friction angle φ. In the triaxial test (Figure 7.6), the theoretical expression of the Mohr-Coulomb is:

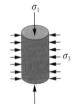

Figure 7.6 Triaxial test of the asphalt mixture (JTG E20—2011 T0718)

$$\sigma_1 = \frac{1 + \sin\varphi}{1 - \sin\varphi}\sigma_3 + 2c\frac{\cos\varphi}{1 - \sin\varphi} \tag{7.1}$$

where σ_1 = the maximum principal stress (kPa);

σ_3 = the minimum principal stress (kPa).

For a given material, parameters c and φ are constant. There is a linear relationship between σ_1 and σ_3 as below.

$$\sigma_1 = k\sigma_3 + b \tag{7.2}$$

where k and b are coefficients.

The parameters c and φ can be calculated as

$$\begin{cases} \sin\varphi = \dfrac{k-1}{k+1} \\ c = \dfrac{b}{2} \cdot \dfrac{1-\sin\varphi}{\cos\varphi} = \dfrac{b}{2\sqrt{k}} \end{cases} \quad (7.3)$$

In the simple tensile-compressive test, the c and φ values of the asphalt mixture can also be obtained based on the unconfined compressive strength R and the tensile strength r. Under unconfined compression, $\sigma_3 = 0$, $\sigma_1 = R$, we have

$$R = \sigma_1 = \dfrac{2c\cos\varphi}{1-\sin\varphi} = 2c \cdot \tan\left(\dfrac{\pi}{4} + \dfrac{\varphi}{2}\right) \quad (7.4)$$

Under tension, $\sigma_1 = 0$, $-\sigma_3 = r$, we have

$$r = -\sigma_1 = \dfrac{2c\cos\varphi}{1+\sin\varphi} = 2c \cdot \cot\left(\dfrac{\pi}{4} + \dfrac{\varphi}{2}\right) \quad (7.5)$$

The parameters c and φ can be calculated as

$$c = \dfrac{1}{2}\sqrt{Rr} \quad (7.6)$$

$$\sin\varphi = \dfrac{R-r}{R-r} \quad (7.7)$$

7.2.3 Influencing Factors

1) Internal factors

(1) Asphalt content

The two curves in Figure 7.7 show the change of the cohesion and internal friction angle by the influence of asphalt content. With the increase of asphalt content, the internal friction angle decreases because of the lubrication of asphalt. However, the cohesion

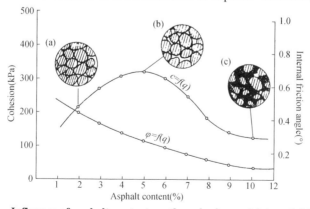

Figure 7.7 Influence of asphalt content on the cohesion and internal friction angle

increases first, and then decreases. This is because when the asphalt content is too low, there is not enough asphalt to coat and bind aggregates together. When the asphalt content is too high, the mixture became too plastic. Therefore, the optimum asphalt content should be determined in the mixture design.

(2) Asphalt viscosity

When the asphalt mixture is subjected to shear action, especially when subjected to dynamic loads, the asphalt with a high viscosity can provide great viscous resistance and high shear strength. With the increase of asphalt viscosity, the cohesion of the asphalt mixture increases significantly, and the internal friction angle also increases slightly.

(3) Aggregates

Large aggregate size increases internal friction, but the cohesion is low. Dense gradation has high cohesion but low internal friction; open gradation has low cohesion but high internal friction; gap gradation has both high cohesion and infernal friction. The rough surface of aggregates also increases the internal friction of the asphalt mixture. The effects of aggregate gradation on the performance of asphalt mixtures are listed in Table 7.2.

Table 7.2 Characteristics of asphalt mixtures with different gradations

Gradation	Dense	Open	Gap
Structures	Suspension-dense	Skeleton-open	Skeleton-dense
Composition	• More fine and medium size particles • Mortar content: Low • Air voids: low	• More coarse and very few fines • Mortar content: Low • Air voids: high	• More coarse and a few fines • Mortar content: high • Air voids: low
Parameters	$c\uparrow, \varphi\downarrow$	$c\downarrow, \varphi\uparrow$	$c\uparrow, \varphi\uparrow$
Properties	• Low permeability • High cohesion • Particle interlocking: Low friction • High crack resistance • Better durability • Poor high temperature stability	• High interconnected air voids • Moisture susceptibility • Low fatigue resistance • Low durability • Need high adhesive asphalt	• High cohesion • High internal friction • Good high- & low-temperature stability • High durability • Cautious for segregation

(4) Interaction between asphalt and aggregates

The interaction between asphalt and aggregates directly influences strength, temperature and moisture susceptibility, and aging. The interaction between asphalt and aggregates is a complex process, which can be divided into physical absorption and chemical absorption. Physical absorption occurs when there is only a molecular force (van der Waals force). Chemical absorption occurs when asphalt and aggregates form compounds. The alkaline aggregates react with asphalt containing a specific amount of acidic surfactant, which can form new compounds on the surface between asphalt and aggregates. Because these compounds are insoluble in water, the asphalt layer formed on the surface of the aggregate material is highly water-resistant. However, for the acidic aggregate which includes more than 65 % SiO_2, chemical absorption compounds will not form and the moisture susceptibility is relatively low.

The interaction between asphalt and mineral materials leads to the redistribution of asphalt chemical components on the surface of mineral materials, producing structural asphalt. The cohesion of structural asphalt is higher than the rest "free" asphalt. For the same amount of asphalt, the larger the surface area of the mineral materials interacting with asphalt, the thinner the asphalt film, and the larger the proportion of structural asphalt in the asphalt, increasing the cohesion of the asphalt mixture. To increase the physical-chemical interaction between asphalt and mineral aggregates, the asphalt mixture must contain an appropriate amount of mineral fillers. Finer fillers can increase the specific area of mineral materials and fillers less than 0.075 mm should be no less than a specific amount. However, the fillers less than 0.005 mm should be strictly controlled; otherwise, it may cause agglomerate in the asphalt mixture during handling.

Excess asphalt content and the increase of the void ratio will increase the amount of free asphalt which will weaken the structural cohesion of the asphalt mixture. However, a certain amount of free asphalt must be presented to ensure the desired corrosion resistance and the optimum ductility of the asphalt mixture.

2) External factors

For external factors, low temperature and high loading rate are related and, in this case, higher cohesion is more important to resist cracking. The asphalt mixture is a visco-elastic material, and its shear strength is closely related to the deformation rate. The deformation rate has little effect on the internal friction angle of the asphalt mixture, but a more significant effect on the cohesion of the asphalt mixture. The experimental results show that the cohesion increases significantly with the increase of the deformation rate, while the internal friction angle changes little. The basic requirements for good-quality asphalt mixtures include: (1) a strong aggregate skeleton, which can be obtained by selecting appropriate gradation and maximizing the interlocking of aggregates; (2) the optimum amount of asphalt for mixing and compaction; and (3) an active mineral of effective chemical absorption with asphalt.

To sum up, the strength of the asphalt mixture can be improved by improving the cohesion or internal friction. To improve cohesion, high viscosity asphalt, rough aggregate surface texture, more fillers, and structural asphalt are preferred. To improve internal friction, aggregates with good angularity, gradation, and surface texture are preferred. Those factors will be considered during the mixture design process, including material selection, gradation design, and optimum asphalt content determination.

7.3 Properties and Tests

To provide sufficient stability during the life cycle service of pavement, the mechanical and physical properties of asphalt mixtures should meet a specific requirement, considering the traffic loads and environmental conditions. The key properties of asphalt mixtures include high-temperature stability, low-temperature cracking, fatigue performance, moisture susceptibility, and friction resistance. Since the asphalt binder is a typical temperature-dependent rheological material, the loading time and temperature control are of significance in evaluating many mechanical properties of asphalt mixtures.

7.3.1 High-Temperature Stability

Typical high temperature distress in asphalt pavements includes bleeding, shoving, and rutting. As shown in Figure 7.8, bleeding occurs in an asphalt pavement when a film of the asphalt binder appears on the pavement surface. Insufficient air voids and excess asphalt are causes of bleeding. Shoving is a form of plastic movement typified by ripples, corrugation, or an abrupt wave or shoving across the pavement surface. The distortion is perpendicular to the traffic direction, and usually occurs in roundabouts where traffic starts and stops frequently. Rutting is pavement surface depression in the wheel path due to traffic loads. Permanent deformation is a key performance parameter that can depend largely on the asphalt mixture design. Therefore, most performance test efforts focus on deformation resistance prediction.

(a) Bleeding　　　　　(b) Shoving　　　　　(c) Rutting

Figure 7.8　High temperature distress in asphalt pavements

Causes for permanent deformation problems include the poor quality of asphalt, aggregates, mixture proportion, insufficient compaction during construction, subgrade deformation due to excessive moisture, and extreme traffic and weather conditions. In general, a fine aggregate asphalt mixture tends to have high plastic deformation than a coarse aggregate asphalt mixture. Mixtures with low design void content also tend to have greater rutting than mixtures with high void content. It should be noted that air voids and volumes between aggregates (VMA) in the asphalt mixture are critical for rutting resistance as well as other properties. For example, high VMA usually means a weak aggregate skeleton and may cause rutting potential, while low VMA usually means fewer binders and may cause poor fatigue cracking resistance. The rutting resistance of asphalt mixtures can be improved by increasing both the cohesion and the internal friction angle. To ensure sufficient cohesion, it is important to use optimum asphalt content and high viscosity asphalt. To improve the internal friction angle, it is recommended to use aggregates with a rough surface and good angularity and be cautious of flaky and elongated particles in aggregates. A 3%–5% design air void content and effective compaction during construction are recommended to help obtain strong aggregate interlocking in asphalt mixtures.

1) Marshall stability test

The Marshall stability test is originally designed to evaluate the strength of the asphalt mixture for determining optimum asphalt content and can be used as an empirical indicator of rutting resistance. Marshall stability test is characterized by simple equipment and easy operation and has been adopted by many countries. Figure 7.9 shows the apparatus for the Marshall stability test, including a Marshall hammer, a water bath tank, and a Marshall stability tester.

Civil Engineering Materials

(a) Marshall hammer   (b) Water bath tank   (c) Marshall stability tester

Figure 7.9 Marshall stability test apparatus

According to specification JTG E20—2011 T0709, the test involves preparing pill specimens using the Marshall hammer. The specimens are immersed in the 60 ℃ water bath for 2 hours. Then, a lateral load is applied on the side of the specimen at a 51 mm/min loading rate to test its strength using the Marshall stability tester. Figure 7.10 shows the load-deformation curve of the Marshall stability test. The Marshall stability is the maximum load and the flow value is the corresponding deformation. For

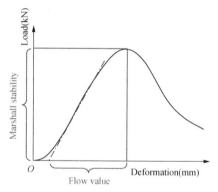

Figure 7.10 Results of the Marshall stability test

the traditional dense-graded HMA, high Marshall stability or low flow value usually indicates high rutting resistance. Marshall stability and flow value are important empirical indicators for asphalt rutting potential and the quality control of asphalt pavement construction. However, many practices and research show that some asphalt mixtures with good Marshall stability and flow value may still have rutting problems. Therefore, many other tests are specifically developed to characterize the rutting problems of asphalt mixtures.

2) Loaded Wheel Test

Many types of laboratory loaded wheel testers are used to measure the permanent deformation by rolling a small loaded wheel device repeatedly across a prepared specimen. According to the specification JTG E20—2011 T0719, the typical wheel-load tester uses a

40 cm × 40 cm × 5 cm slab specimen tested at 60 ℃ for 2 hours. Rutting depth and dynamic stability (DS) which are the number of repeated loadings required to create 1 mm rutting depth are used to evaluate the rutting resistance of the mixture. DS is calculated as

$$DS = \frac{(60-45) \times 42}{D_{60} - D_{45}} \cdot C_1 \cdot C_2 \quad (7.8)$$

where D_{60} = deformation at 60th minute (mm);
D_{45} = the deformation at 45th minute (mm);
C_1, C_2 = coefficients.

3) Hamburg wheel tracking device (HWTD) test

The HWTD, developed in Germany, can be used to evaluate rutting and stripping potential. The HWTD tracks a loaded steel wheel back and forth directly on a specimen. According to the specification AASHTO T 324-14, tests are typically conducted on 260 mm × 320 mm × 40 mm slabs compacted to 7% air voids with a linear kneading compactor. Figure 7.11 shows a typical rut depth versus wheel passes curve from an HWTD test and the key plot parameters. The following parameters are measured and reported:

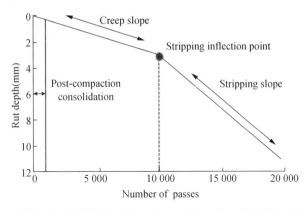

Figure 7.11 An HWTD test showing rutting after 10,000 load cycles

(1) Post-compaction consolidation is the initial consolidation of the mixture. The rut depth at 1000 load cycles is assumed due to continued consolidation.

(2) The creep slope is the inverse of the rutting slope after post-compaction consolidation but before the stripping inflection point. The creep slope is used to evaluate rutting potential instead of rut depth because the number of load cycles at which moisture damage begins to affect rut depth varies between HMA mixtures and cannot be conclusively determined by the plot.

(3) The stripping inflection point is the point at which the creep slope and strip-

ping slope intercept each other, indicating the moisture damage potential. If the stripping inflection point occurs at a low number of load cycles, (e. g. , less than 10000) the mixture may be susceptible to moisture damage.

(4) The stripping slope measures the accumulation of moisture damage.

4) Asphalt pavement analyzer (APA) test

The APA operates like the HWTD but uses different equipment. According to specification AASHTO T 340—10, it can conduct different types of rutting tests and use both beam and cylinder specimens (Figure 7.12). The APA uses an aluminum wheel that is loaded onto a pressurized linear hose and tracked back and forth over the asphalt pavement specimen. The amount of rutting is measured after the wheel has been tracked for a set number of cycles at a constant load and hose pressure. The deformation on the wheel track is recorded during the testing to evaluate the rutting performance.

Figure 7.12 Asphalt pavement analyzer

5) Indirect tensile resilient modulus test

Laboratory wheel-tracking tests can correlate well with in-service pavement rutting. However, these tests are simulative tests and do not measure any fundamental material parameter. Therefore, a more fundamental characteristic of asphalt mixtures for evaluating the structure responses is developed. The specification AASHTO T322-07 recommends an indirect tensile resilient modulus test. It applies pulse load along the vertical diameter of a pill specimen. As shown in Figure 7.13, the indirect tensile resilient modulus test is a

Figure 7.13 Indirect tensile resilient modulus test

fatigue test with a short period and low cost. It is the repeated compression load acting parallel to the vertical radial surface of the cylindrical specimen. This type of loading method produces uniform tensile stress along the vertical radial surface and perpendicular to the direction of load action. The test is easy to operate and is adopted by the majority of researchers.

The diameter of the specimen is 100 mm, the height is 63.5 mm, and the load acts on the specimen through a strip with a width of 12.5 mm. The load is usually applied with a duration of 0.1 s and a rest period of 0.9 s. The recoverable horizontal deformation is also measured during the test. Figure 7.14 shows the typical load and horizontal deformation versus time relationships. The test is commonly performed at three temperatures: 5 ℃, 25 ℃, and 40 ℃, and the resilient modulus, which describes the stress-strain relationship of asphalt mixtures, can be calculated as

$$M_R = \frac{P(0.27 + v)}{t\Delta H} \tag{7.9}$$

where P = applied load (N);
M_R = resilient modulus (MPa);
t = thickness of the specimen (mm);
v = Poisson's ratio;
ΔH = recoverable horizontal deformation (mm).

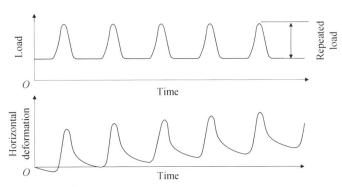

Figure 7.14 Repeated load and horizontal deformation in the indirect tensile resilient modulus test

6) Asphalt mixture performance test (AMPT)

In addition to the new PG binder testing system, the Superpave program also develops the AMPT or formerly known as simple performance tester (SPT), which is designed to evaluate the viscoelastic behavior of asphalt concrete. According to specification

AASHTO T378-17, all these tests use 100 mm in diameter and 150 mm in height cylindrical specimens (Figure 7.15). Specimens are cored from pills compacted using the Superpave gyratory compactor. The AMPT is the companion performance test for use with the Superpave mix design procedure.

Figure 7.15 AMPT equipment and the specimen

As summarized in Table 7.3, the tester can conduct three tests: the dynamic modulus test, the triaxial static creep test, and the triaxial repeated load permanent deformation test. From the three tests, three parameters can be obtained. The dynamic modulus is to evaluate both permanent deformation and fatigue cracking, flow time and flow number are to evaluate permanent deformation.

Table 7.3 Asphalt Mixture Performance Tests

Tests	Parameters	Permanent deformation	Fatigue cracking
Dynamic modulus test	Dynamic modulus	✓	✓
Triaxial static creep test	Flow time	✓	
Triaxial repeated load permanent deformation test	Flow number	✓	

(1) Dynamic modulus test

According to specification AASHTO T342-11, the dynamic modulus test consists of applying an axial sinusoidal compressive stress to an unconfined or confined specimen as shown in Figure 7.16. Assuming that the asphalt mixture is a linear viscoelastic material, a complex number called the complex modulus can be obtained from the test to define the relationship between stress and strain. Test specimens can be tested at different temperatures and three different loading frequencies (commonly 1, 4, and 16 Hz). The applied load varies and is usually applied in a haversine wave. The dynamic modulus is indicative of the stiffness of the asphalt mixture at the selected temperature and load frequency. The dynamic modulus is correlated to both rutting and fatigue cracking of the mixture and is required in the mechanistic-empirical pavement design guide (MEPDG).

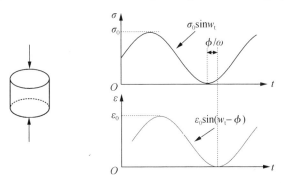

Figure 7.16 Stress-strain relationship under a continuous sinusoidal loading

As shown in Figure 7.16, for linear viscoelastic materials, in which the stress-strain ratio is independent of the applied loading stress, this relationship is defined by a complex number called the "complex modulus" (E^*) Equation (7.10).

$$E^* = E' + iE'' = |E^*|\cos\varphi + i|E^*|\sin\varphi \tag{7.10}$$

where E^* = complex modulus;

E' = storage or elastic modulus;

E'' = loss or viscous modulus;

$|E^*|$ = dynamic modulus;

φ = phase angle;

i = imaginary number.

Similar to the complex modulus, G^*, in the DSR test of the asphalt binder, the real and imaginary portions of the dynamic modulus are the elastic modulus and the viscous modulus. The dynamic modulus is the absolute value of the complex modulus, defined as the peak dynamic stress divided by the peak recoverable axial strain, and can be calculated by Equation (7.11).

$$|E^*| = \frac{\sigma_0}{\varepsilon_0} \tag{7.11}$$

where σ_0 = peak amplitude of stress;

ε_0 = peak amplitude of recoverable axial strain.

The phase angle φ is the angle by which strain lags behind stress. It is an indicator of viscous properties and is calculated by Equation (7.12).

$$\varphi = \frac{t_i}{t_p} \times 360 \qquad (7.12)$$

where t_i = time lag between stress and strain cycles;

t_p = time for a stress cycle.

(2) Triaxial static creep test

According to specification GB/T 229—2020, the triaxial static creep test uses either one load-unload cycle or incremental load-unload cycles. It provides information to determine the instantaneous elastic and plastic components, or the viscoelastic and viscoplastic components of the material's response. Both of the two components are time-dependent. It uses a confining pressure of about 138 kPa, which allows test conditions to match field conditions more closely. From the test, the compliance, $D(t)$, which is strain as a function of time $\varepsilon(t)$ over constant stress σ_0 can be calculated by Equation (7.13).

$$D(t) = \frac{\varepsilon(t)}{\sigma_0} \qquad (7.13)$$

where $\varepsilon(t)$ = strain as a function of time;

σ_0 = constant stress.

As shown in Figure 7.17, the compliance can be divided into three major zones: primary zone, secondary zone, and tertiary flow zone. The primary zone is the initial consolidation, the secondary zone is the slow accumulation of deformation, and the tertiary flow zone is the fast deformation indicating a potential shear failure. The time at which tertiary flow starts is the flow time and is usually used to evaluate the rutting resistance of the asphalt mixture.

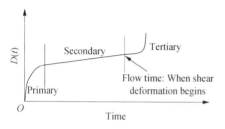

Figure 7.17 Three stages of triaxial static creep test

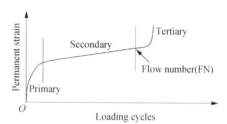

Figure 7.18 Three stages of triaxial repeated load permanent deformation test

(3) Triaxial repeated load permanent deformation test

The triaxial repeated load permanent deformation test records the cumulative permanent deformation of the mixture as a function of the number of load repetitions. The test applies a repeated load of fixed magnitude and cycle duration to a cylindrical test specimen. The specimen's resilient modulus can be calculated using the horizontal deformation and an assumed Poisson's ratio. For the triaxial repeated load permanent deformation test, the loading mode is haversine loading with 0.1 second loading and 0.9 second resting. Similar to the compliance in the triaxial static test, the cumulative permanent strain curve in the triaxial repeated load permanent deformation test can also be divided into three zones: primary, secondary, and tertiary (Figure 7.18). The cycle number at which tertiary flow starts is the flow number (FN).

7.3.2 Low-Temperature Cracking

Typical distress of asphalt pavement at low temperatures is thermal cracking, mostly appearing as transverse cracking or longitudinal cracking (Figure 7.19). As the temperature drops, the restrained pavement tries to shrink. The tensile stress builds up to a critical point at which a crack is formed. Thermal cracks can be initiated by a single low-temperature event or by multiple warming and cooling cycles and then propagated by further low temperatures. For the mixture, the shrinkage, stiffness modulus, fracture strain, and tensile strength at low temperatures all influence the resistance to thermal cracking. The bending beam test and direct tension test in the PG grade asphalt binder tests are for thermal cracking. The tensile strength of the mix is related to pavement cracking, especially at low temperatures. The tensile strength is important because it is a good indicator of cracking potential. A high tensile strain at failure indicates that a particular mixture can tolerate a higher strain before failing, which means it is more likely to resist cracking than asphalt with a low tensile strain at failure.

Figure 7.19 Thermal cracking on asphalt pavements

Civil Engineering Materials

There are two tests typically used to measure tensile strength: the indirect tensile test (IDT) and the semi-circle bending test (SCB) as shown in Figure 7.20. They can be conducted at either intermedium or low temperatures such as 15 or $-10\ \text{℃}$. The IDT and SCB tensile strength can be calculated by Equations (7.14) and (7.15), respectively. In addition to the tensile strength, the fracture energy or toughness, defined as the area bounded by the force-displacement curve can also be used to evaluate the cracking resistance of the mixture.

$$T_i = \frac{2F}{\pi bD} \tag{7.14}$$

$$T_s = \frac{6FL}{bD^2} \tag{7.15}$$

where T_i = IDT tensile strength (MPa);
F = ultimate load (N);
b = thickness of the specimen (mm);
D = diameter of the specimen (mm);
T_s = SCB tensile strength (MPa);
L = span between two supporting rollers (mm).

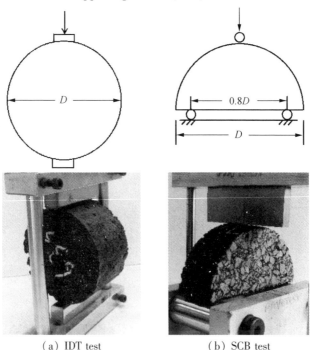

(a) IDT test (b) SCB test
Figure 7.20 IDT and SCB Tests

7   Asphalt Mixtures

7.3.3   Fatigue Performance

Unlike rutting and thermal cracking which are related to high and low temperatures respectively, fatigue damage occurs at an intermedium temperature. Typical fatigue cracking appears as a network of interconnected cracks (Figure 7.21). Frost action and edge failure can also cause fatigue cracking. Fatigue is caused by the damage accumulation under repeated action of loads lower than strength. The repeated stress value of asphalt mixtures with fatigue failure is usually called the fatigue strength and the corresponding number of repeated stress actions is called the fatigue life. As shown in Figure 7.22, there are two mechanisms of fatigue cracking in asphalt pavements. In thin pavements, cracking initiates at the bottom of the HMA layer where the tensile stress is the highest, then it propagates to the surface as one or more longitudinal cracks. This is commonly referred to as "bottom-up" fatigue cracking. In thick pavements, the cracks most likely initiate from the top in areas of high localized tensile stress resulting from the tire-pavement interaction and asphalt binder aging. This is commonly referred to as "top-down"

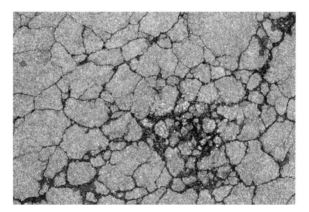

Figure 7.21   Fatigue cracking in asphalt pavements

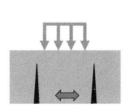
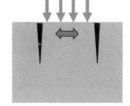

(a) Bottom-up fatigue cracking   (b) Top-down fatigue cracking

Figure 7.22   Two types of fatigue cracks in asphalt pavements

fatigue cracking. The "bottom-up" fatigue cracking usually occurs on the sides of the longitudinal wheel paths and may develop into interconnected fatigue cracks.

The flexural fatigue test, also called the four-point beam bending test, is usually conducted to evaluate fatigue performance. According to specification JTG E20—2011 T0739, the fatigue life of a small beam specimen is determined by subjecting it to repeated flexural bending until failure. It usually uses the strain control mode. Stiffness is measured at the 50th load cycle, and failure is determined when the initial stiffness is reduced by 50%. The number of loading cycles to failure indicates fatigue life and dissipated energy measures the damage to the specimen. Dissipated energy is calculated based on the stress and strain relationship. Dissipated energy is a measure of the energy that is lost in the material or altered through mechanical work, heat generation, or damage to the specimen.

As shown in Figure 7.23 and Figure 7.24, the 40 mm × 40 mm × 380 mm beam is held in place by four clamps and a repeated haversine load is applied to the two inner clamps with the outer clamps providing a reaction load. The loading rate is normally set between 1 Hz to 10 Hz. This setup produces a constant bending moment over the center portion of the beam. The deflection caused by the load is measured at the center of the beam. The load repetitions to failure are calculated by Equation (7.16):

$$N_f = c\left(\frac{1}{\varepsilon}\right)^m \tag{7.16}$$

where N_f = repeated loading times to fatigue failure;

ε = applied strain level;

c, m = regression coefficients related to the properties of asphalt mixtures.

Figure 7.23　Flexural fatigue test device and the beam specimen

7 Asphalt Mixtures

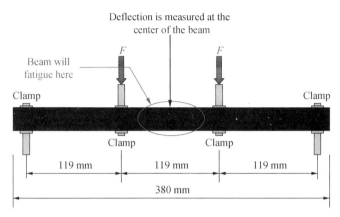

Figure 7.24 Flexural fatigue test method

Based on the flexural fatigue test, we can obtain the strain level versus fatigue life equation (Equation 7.16) and the curve (Figure 7.25). In the strain level vs. fatigue life curve, the horizontal axis of fatigue life uses logarithm transformation due to the large values. Generally, the higher the strain level, the shorter the fatigue life.

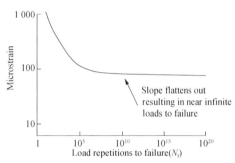

Figure 7.25 Typical flexural fatigue test

In mechanics, a fatigue limit or an endurance limit is the stress level below which an infinite number of loading cycles can be applied to a material without causing fatigue failure. Recent studies suggest that at low levels of strain, around 70 microstrains, asphalt mixtures have, in effect, an infinite fatigue life. This is because a continuous physical-chemical healing reaction occurs at low strain levels. The healing potential is that the asphalt mixture can recover some constant amount of damage. If the damage due to loading falls below this "healing potential", the damage accumulation is virtually non-existent. Based on this theory, the concept of "perpetual pavement" is developed. Researchers believe that a pavement can have infinite life if it is thick enough to only have a very small strain.

7.3.4 Moisture Susceptibility

Moisture damage to asphalt pavements includes stripping of asphalt, aggregate raveling, and potholes. Stripping of asphalt is the loss of adhesion between asphalt and aggregates, mostly due to moisture damage. Raveling occurs as individual aggregate particles dislodge from the pavement surface (Figure 7.26). It usually starts with the loss of fine aggregates and advances to the loss of larger aggregates. A pothole is a bowl-shaped hole in an asphalt pavement surface that usually forms as a result of water seeping into pavement cracks and freezing during the winter season.

Figure 7.26 Moisture damage to asphalt pavements

The HWTD test results give a relative indication of the moisture susceptibility of the mixture. Many performance tests that can be conducted on a conditioned specimen can be used to evaluate the moisture susceptibility by comparing the test results of conditioned and unconditioned specimens. For example, measuring tensile strength before and after water conditioning can give some indication of the moisture susceptibility. If the water-conditioned tensile strength is relatively high compared to the dry tensile strength, then the mixture can be assumed reasonably moisture resistant.

The freeze-thaw test is the most widely used test to evaluate a mixture's resistance to moisture damage. It compares the indirect tensile strength test results of a dry specimen and a specimen exposed to freezing-thawing. According to specifications AASHTO T283-07 and JTG E20—2011 T0729, the specimens are saturated with water to 50%–80% using a vacuum machine, frozen at −18 ℃ for 16 hours, and then immersed in 60 ℃ water for 24 hours before tests (Figure 7.27). The tensile strength ratio (TSR) is calculated by Equation (7.17). A TSR higher than 80% is required, indicating the strength loss due to one freeze-thaw cycle is less than 20%.

$$TSR = \frac{IDT_{\text{after}}}{IDT_{\text{before}}} \times 100 \qquad (7.17)$$

where IDT_{after} = average tensile strength of conditioned specimens (MPa);

IDT_{before} = average tensile strength of unconditioned specimens (MPa).

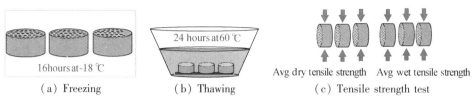

Figure 7.27 Procedure of freeze-thaw test

Similarly, we can use the retained Marshall stability to evaluate the moisture susceptibility. The Marshall immersion test includes conditioning the saturated specimens in 60 ℃ water for 48 hours. The other set of specimens is immersed in water for 30 to 40 minutes. Then, a Marshall tester is used to test the stability and flow values. The retained Marshall stability is calculated by

$$MS_r = \frac{MS_{after}}{MS_{before}} \times 100 \tag{7.18}$$

where MS_r = retained Marshall stability (%);

MS_{after} = Marshall stability of conditioned specimens (kN);

MS_{before} = Marshall stability of unconditioned specimens (kN).

7.3.5 Friction

The pavement surface should provide sufficient friction for traffic safety concern. The friction is mainly determined by pavement surface texture and aggregate angularity. Surface friction is related to the pavement surface's micro texture and macro texture. The micro texture refers to the small-scale texture of the pavement aggregate components, while the macro texture refers to the large-scale texture of the pavement as a whole due to the aggregate particle arrangement. Angular materials are desirable in paving mixtures because they tend to lock together and resist deformation after initial compaction, whereas rounded materials may not produce sufficient inter-particle friction to prevent rutting.

Figure 7.28 illustrates the two frequently used tests for the surface friction of the asphalt mixture. The sand patch test (JTG E20—2011 T0731) is carried out on a dry pavement surface by pouring a known quantity of sand onto the surface and spreading it in a circular pattern. As the sand is spread, it fills the low spots on the pavement surface. When the sand cannot be spread any further, the diameter of the resulting circle is measured. This diameter can then be correlated to an average texture depth, which can be correlated to friction.

Civil Engineering Materials

(a) Sand patch test

(b) Pendulum skid tester

Figure 7.28 Friction tests of the asphalt mixture

The pendulum skid tester (JTG E60—2008 T0964) can also be used to measure pavement surface friction. The highest point of the swing pendulum indicates the skid resistance. The higher the swing, the lower the skid resistance. The pavement surface temperature influences the measured friction coefficient at lower temperatures. With the decrease of temperature, the friction decreases. When the pavement surface temperature is above 40 ℃, the temperature change has little effect.

To ensure sufficient pavement surface friction, it is recommended to use aggregates with good angularity and abrasion resistance. The asphalt content should also be controlled to prevent excess asphalt, bleeding, and friction loss.

7.4 Volumetric Parameters

7.4.1 Volumes in Mixture

The volumetric parameters of the asphalt mixture are fundamental for asphalt mixture design. Regardless of the method used, the mix design is a process to determine the volume of asphalt and aggregates required to produce a mixture with the desired properties. As shown in Figure 7.29, in an asphalt mixture, we firstly have the aggregate and voids in mineral aggregate (VMA). When asphalt is added, part of the asphalt is absorbed by the aggregate, filling a portion of the permeable air voids in the aggregate. The unabsorbed asphalt is called effective asphalt and it partially fills VMA. The unfilled voids in the VMA are the air voids. Figure 7.29 shows the weight and volume proportions in an asphalt mixture. The three key volumetric parameters for the asphalt mixture are discussed below.

7 Asphalt Mixtures

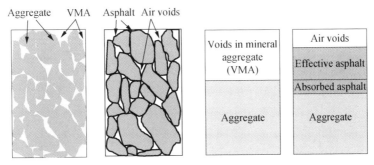

Figure 7.29 Volumes in the mineral aggregate and mixture

1) Voids in total mix (VTM)

VTM is the voids in the total mix (Figure 7.29). It is the ratio of the volume of air voids over the bulk volume of the mixture as shown in Figure 7.30 and Equation 7.19. A specific percentage of air voids is necessary for all dense-graded highway mixtures to allow for some additional pavement compaction under traffic and to provide space in which small amounts of asphalt can flow during this subsequent compaction. Insufficient voids in a mixture may cause aggregate particles to lose contact with each other due to asphalt expansion at elevated temperatures, resulting in a loss of stability and increased potential for rutting under traffic loads. High voids, however, can decrease the durability of a pavement by allowing water and air to permeate the mix, increasing the oxidization and stripping potential, resulting in reduced mixture durability. Voids also contribute to the thermal stability of compacted asphalt concrete by allowing for thermal expansion of binders between the aggregate particles as well as volumetric strain under repeated heavy traffic loading.

$$VTM = \frac{V_a}{V_m} \times 100 \qquad (7.19)$$

where *VTM* = percentage of voids in total mix (%);

V_a = volume of air voids;

V_m = bulk volume of the mixture.

2) Voids in mineral aggregate (VMA)

VMA is the voids in the mineral aggregate (Figure 7.29). It is the ratio of the volume of air voids and effective asphalt over the bulk volume of the mixture as shown in Figure 7.30 and Equation (7.20). The *VMA* represents the space that is available to accommodate the asphalt and the volume of air voids necessary in the mixture. The more *VMA* in the dry aggregate, the more space is available for the film of the asphalt binder. *VMA* is critical to the rutting and fatigue resistance of the mixture. High *VMA* usually means a weak aggregate skeleton and may cause rutting potential, while low *VMA* usually means fewer binders and may cause poor fatigue cracking resistance.

$$VMA = \frac{V_a + V_{be}}{V_m} \times 100 \qquad (7.20)$$

where *VMA* = percentage of voids in mineral aggregates (%);

V_{be} = volume of effective asphalt binder.

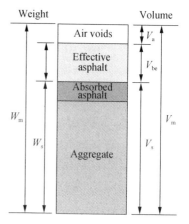

Figure 7.30 Weights and volumes in the asphalt mixture

3) Voids filled with asphalt (VFA)

VFA is the *VMA* filled with effective asphalt (Figure 7.29). It is the ratio of the volume of effective asphalt over the volume of air voids and effective asphalt as shown in Figure 7.30 and Equation (7.21). The *VFA* is an important measure of relative durability. If the *VFA* is too low, there are insufficient binders to provide durability and to over-densify under traffic and bleeding.

$$VFA = \frac{V_{be}}{V_a + V_{be}} \times 100 \qquad (7.21)$$

where *VFA* = percentage of voids filled with asphalt (%).

Figure 7.31 shows the details of volumes and weights in an asphalt mixture. We usually record different volumetric parameters of the asphalt mixture in the format of Z_{xy}. Z can be V, W, G, and P, representing volume, weight, specific gravity, and percentage, respectively; x can be a, b, s, and m, representing air, binder, stone or aggregate, and mixture, respectively; y can be b, e, a and m, representing bulk, effective, apparent/absorbed and maximum, respectively. For example, V_{ba} means volume, binder and absorbed, and therefore it is the volume of absorbed asphalt binder. G_b means the specific gravity of asphalt binder. G_{se} means specific gravity, stone, and effective, and it is the effective specific gravity of the aggregate. G_{mm} means specific gravity, mixture, and maximum, and it is the theoretical maximum specific gravity of the mixture.

7 Asphalt Mixtures

$$Z_{xy} \quad Z: \begin{matrix} V = \text{volume} \\ W = \text{weight} \\ G = \text{specific gravity} \\ P = \text{percent} \end{matrix} \quad x: \begin{matrix} a = \text{air} \\ b = \text{binder} \\ s = \text{stone} \\ m = \text{mixture} \end{matrix} \quad y: \begin{matrix} b = \text{bulk} \\ e = \text{effective} \\ a = \text{apparent/absorbed} \\ m = \text{maximum} \end{matrix}$$

V_T = total volume of the compacted specimen;
V_a = volume of air voids;
V_b = volume of asphalt binder;
V_{be} = volume of effective asphalt binder;
V_{ba} = volume of absorbed asphalt binder;
V_{sb} = bulk volume of the aggregate;
V_{se} = effective volume of the aggregate = $(V_t - V_a - V_b)$

W_T = total weight of the compacted specimen
W_D = dry weight
W_{SSD} = saturated surface dry (SSD) weight
W_{sub} = weight submerged in water
W_b = weight of the asphalt binder
W_{be} = weight of effective asphalt binder
W_{ba} = weight of absorbed asphalt binder
W_s = weight of the aggregate

G_{sa} = apparent specific gravity of aggregates;
G_b = specific gravity of asphalt binder;
G_{sb} = bulk specific gravity of aggregates;
G_{se} = effective specific gravity of aggregates;
G_{mb} = bulk specific gravity of the compacted mixture;
G_{mm} = theoretical maximum specific gravity of the mixture

P_b = asphalt content by weight of mixture (%);
P_s = aggregate content by weight of mixture (%);
P_a = content of air voids (%);
P_{be} = effective asphalt content by weight of mixture (%)

γ_w = unit weight of water

Figure 7.31 Volumetric parameters in the asphalt mixture

7.4.2 Calculation of Parameters

1) Bulk specific gravity of the compacted mixture

The bulk specific gravity of the compacted mixture is measured similarly to the coarse aggregate. We first measure the dry weight of the specimen, suspend the specimen from a scale into water for 3 to 5 minutes and obtain the submerged weight, and then dry the specimen surface with a moist towel to remove water at the surface without removing water from the surface voids of the specimen and measure the saturated surface dry weight (Figure 7.32). The bulk specific gravity of the compacted mixture can be calculated by Equation (7.22).

$$G_{mb} = \frac{W_d}{W_{SSD} - W_{sub}} \tag{7.22}$$

where G_{mb} = bulk specific gravity of the compacted mixture;

W_D = specimen dry weight in the air (g);

W_{sub} = specimen weight in water (g);

W_{SSD} = specimen saturated surface dry weight (g).

Civil Engineering Materials

Figure 7.32 Bulk specific gravity test

2) Theoretical maximum specific gravity

To measure the theoretical maximum specific gravity, we have to obtain the particles of the mix or the "rice sample" by separating the mix into particles and then using a vacuum to remove all trapped air (Figure 7.33). The theoretical maximum specific gravity of the mix is the apparent specific gravity of the rice sample. It can be obtained in the same way as that of the coarse aggregate. After measuring the sample dry weight, the bowl weight in water, and the sample and bowl weight in water, the theoretical maximum specific gravity can be calculated by Equation (7.23).

$$G_{mm} = \frac{W_{dr}}{W_{dr} + W_{bowl} - W_{subm}} \tag{7.23}$$

where G_{mm} = theoretical maximum specific gravity of the mix;

W_{dr} = sample dry weight in the air (g);

W_{bowl} = bowl weight in water (g);

W_{subm} = sample and bowl weight in water (g).

Figure 7.33 Theoretical maximum specific gravity test

3) Effective specific gravity of aggregates

It is very difficult to measure the effective specific gravity of aggregates. Instead, it is usually calculated based on the theoretical maximum specific gravity of the mix. We already know that G_{mm} is the weight of the aggregate and binder over the effective volume of the aggregate and volume of the binder as shown in Equation (7.24). The effective volume of the aggregate is the weight of the aggregate over its effective specific gravity G_{se}. Therefore, we can calculate G_{se} based on G_{mm} by Equation (7.25). During mix design, we usually measure the G_{mm} at one asphalt content and calculate the G_{se}. Then, the G_{mm} at the other asphalt content can be easily determined.

$$G_{mm} = \frac{100}{\frac{P_s}{G_{se}} + \frac{P_b}{G_b}} \tag{7.24}$$

$$G_{se} = \frac{P_s}{\frac{100}{G_{mm}} - \frac{P_b}{G_b}} \tag{7.25}$$

where G_{mm} = theoretical maximum specific gravity of the mix at asphalt content $P_b(\%)$;
P_b = asphalt content (%);
P_s = aggregate content, $P_s = 100 - P_b(\%)$;
G_b = asphalt specific gravity at 25 ℃;
G_{se} = aggregates' effective specific gravity.

4) VTM

The *VTM* is the ratio of the volume of air over the bulk volume of the mixture. By substituting the volumes with the dry weight over different specific gravities, the VTM can be calculated as below.

$$VTM = \left(1 - \frac{G_{mb}}{G_{mm}}\right) \times 100 \tag{7.26}$$

where G_{mb} = bulk specific gravity of the mixture;
G_{mm} = theoretical maximum specific gravity of the mixture.

5) VMA

The *VMA* is the difference between the bulk volumes of the mixture and aggregates over the bulk volume of the mixture. By substituting the volumes with the dry weight over different specific gravities, we can obtain the calculation equation for VMA as below.

$$VMA = 100 - \frac{G_{mb}}{G_{sb}} P_s \tag{7.27}$$

where P_s = percentage of aggregates in the asphalt mixture (%);

G_{sb} = aggregates' bulk specific gravity.

6) VFA

The VFA can be calculated as the difference between VMA and VTM over VMA.

$$VFA = \frac{VMA - VTM}{VMA} \times 100 \tag{7.28}$$

7) **Dust to effective binder ratio**

The dust to effective binder ratio is also related to the performance of the asphalt mixture. It is the percentage of aggregates passing the 0.075 mm sieve divided by the effective asphalt content. To calculate this, we have to obtain the percentages of effective and absorbed asphalt. The percentage of absorbed asphalt, P_{ba}, is the weight of absorbed asphalt over the weight of aggregates. Since the volume of absorbed asphalt can be estimated based on the bulk and the effective specific gravity of aggregates, P_{ba} can be calculated based on G_{sb}, G_{se}, and G_b as shown in Equation (7.29). Then, the percentage of effective asphalt content in the mix, P_{be}, and the dust to binder ratio, D/B, can be calculated by Equations (7.30) and (7.31).

$$P_{ba} = \frac{V_{be} G_b}{W_s} = \frac{(V_{sb} - V_{se}) G_b}{W_s} = \frac{\left(\frac{W_s}{G_{sb}} - \frac{W_s}{G_{se}}\right) G_b}{W_s} = 100 \left(\frac{1}{G_{sb}} - \frac{1}{G_{se}}\right) G_b \tag{7.29}$$

$$P_{be} = P_b - \left(\frac{P_{ba}}{100}\right) P_s \tag{7.30}$$

$$D/B = \frac{P_D}{P_{be}} \tag{7.31}$$

where P_{ba} = percentage of absorbed binder based on the weight of aggregates (%);

P_{be} = percentage of effective binder content in the mixture (%);

G_{sb} = bulk specific gravity of the aggregate;

G_{se} = effective specific gravity of the aggregate;

G_b = specific gravity of asphalt;

P_D = percentage of dust or aggregates passing the 0.075 mm sieve (%);

D/B = dust to binder ratio.

An example is presented below to help understand those calculations. We have the following information. We need to calculate voids in the total mix VTM, voids in the mineral aggregate VMA, voids filled with asphalt VFA, effective specific gravity of the aggregate G_{se}, effective binder content P_{be} and theoretical maximum specific gravity of

the mix G_{mm} at 4% asphalt content.
- Bulk specific gravity of the mix $G_{mb} = 2.442$
- Theoretical maximum specific gravity of the mix $G_{mm} = 2.535$
- Bulk specific gravity of the aggregate, $G_{sb1} = 2.716$, $G_{sb2} = 2.689$
- Aggregate content, $P_1 = 50\%$, $P_2 = 50\%$
- Asphalt specific gravity, $G_b = 1.03$
- Asphalt content $P_b = 5.3\%$

Answers:

1) Firstly, we can calculate the bulk specific gravity of the aggregate, G_{sb} (rounding to three decimal places).

$$G_{sb} = \frac{\sum_{i=1}^{n} P_i}{\sum_{i=1}^{n} \frac{P_i}{G_{sbi}}} = \frac{50.0 + 50.0}{\frac{50.0}{2.716} + \frac{50.0}{2.689}} = 2.702$$

2) VTM is calculated based on G_{mm} and G_{mb} (rounding to one decimal place).

$$VTM = 100 \times \frac{G_{mm} - G_{mb}}{G_{mm}} = 100 \times \frac{2.535 - 2.442}{2.535} = 3.7$$

3) VMA is calculated based on G_{mb}, G_{sb} and P_s (rounding to one decimal place).

$$VMA = 100 - \frac{G_{mb}}{G_{sb}} P_s = 100 - \frac{2.442 \times 94.7}{2.702} = 14.4$$

4) VFA can be calculated based on VTM and VMA (rounding to one decimal place).

$$VFA = 100 \times \frac{VMA - VTM}{VMA} = 100 \times \frac{14.4 - 3.7}{14.4} = 74.3$$

5) G_{se} can be calculated based on P_b, G_b and G_{mm} (rounding to three decimal places).

$$G_{se} = \frac{100 - P_b}{\frac{100}{G_{mm}} - \frac{P_b}{G_b}} = \frac{100 - 5.3}{\frac{100}{2.535} - \frac{5.3}{1.030}} = 2.761$$

6) P_{ba} can be calculated based on G_{sb}, G_{se} and G_b (rounding to one decimal place).

$$P_{ba} = 100 \left(\frac{1}{G_{sb}} - \frac{1}{G_{se}} \right) G_b = 100 \times \left(\frac{1}{2.702} - \frac{1}{2.761} \right) \times 1.03 = 0.8$$

7) P_{be} can be calculated based on P_{ba} and P_b (rounding to one decimal place).

$$P_{be} = P_b - \left(\frac{P_{ba}}{100} \right) P_s = 5.3 - \left(\frac{0.8}{100} \right) \times (100 - 5.3) = 4.5$$

8) With G_{se}, the G_{mm} at 4% asphalt content is (rounding to three decimal places).

$$G_{mm} = \frac{100}{\dfrac{P_s}{G_{se}} + \dfrac{P_b}{G_b}} = \frac{100}{\dfrac{100-4}{2.761} + \dfrac{4}{1.030}} = 2.587$$

7.5 Marshall Mix Design

Asphalt mix design is to determine the proportions of asphalt binder and aggregates required to produce a mixture with the desired properties. The main goal of the bituminous mixture design is to achieve the appropriate quantity of bitumen to ensure an adequate coating of aggregates and to provide good workability, resistance to deformation and cracking, and durability of the mixture. In the case of surface courses, asphalt mixtures also should provide the adequate texture and skid resistance for the pavement surface to ensure the passage of vehicles with maximum levels of comfort and safety. Typical asphalt mix design is essential to enhance or mitigate the various performance issues that are to be addressed and these issues include: (1) resistance to permanent deformation, (2) resistance to fatigue and reflective cracking, (3) resistance to low-temperature thermal cracking, (4) durability, (5) resistance to moisture damage, (6) workability, and (7) skid resistance.

7.5.1 Background

During world war II, the US Army Corps of Engineers was looking for a method to design a qualified airport pavement asphalt mixture for large military aircraft. They adopted the method developed by Bruce Marshall from the Mississippi Highway Department. It is now one of the world's most widely used methods for asphalt mix design. The objective of the Marshall method is to determine optimum mix proportion with the best pavement performance. The developed apparatus includes the Marshall hammer, the Marshall mold, and the Marshall stability tester.

In specification JTG F40—2004, the whole mix design includes three stages. Stage I is lab mix design, which is to use the Marshall method to select the mix type and materials, determine the aggregate size, gradation, and asphalt content and evaluate mix performance. Stage II is the plant mix design, which aims to determine aggregate bin portions and asphalt content in the plant. With the formula from the lab mix design, we can produce trials in the plant using ±0.3% asphalt content and conduct the Marshall stability test again to obtain the adjusted mix formula. Stage Ⅲ is field validation, which aims to produce a plant mixture for field trials and evaluate the performance of a plant mixture to determine the final mix formula for the plant. The lab mix design includes seven steps which will be discussed in detail below.

(1) Mixture type selection;
(2) Material selection;
(3) Gradation design;
(4) Specimen preparation and specific gravity tests;
(5) Marshall stability test;
(6) Determination of optimum asphalt content;
(7) Performance evaluation.

7.5.2 Procedures

1) Mixture type selection

The first step is to select the asphalt mix type based on the road class and traffic level. Table 7.1 lists the recommended mix types from specification JTG F40—2004. When the mix type is selected, the recommended mix gradation can be determined.

2) Material selection

The second step is to select the asphalt grade based on the road class, climatic region, pavement layer, construction consideration, etc. and to select aggregates satisfying mechanical and physical properties including angularity, crushing values, abrasion, etc.

3) Gradation design

The third step is gradation design, which is to determine the proportions of aggregates based on the gradation of the selected mix type. The graphical or numerical method can be used to design the gradation. Through the evaluation of the performance of different experimental design grading curves, a design target grading curve that can be used in real engineering is finally determined. Table 7.4 presents the gradation requirement for dense-graded asphalt mixture from specification JTG F40—2004.

Table 7.4 Recommended gradation for dense-graded asphalt mixture

Sieve size (mm)	Passing (%)					
	Coarse	Medium		Fine		Sand
	AC-25	AC-20	AC-16	AC-13	AC-10	AC-5
31.5	100	—	—	—	—	—
26.5	90 – 100	100	—	—	—	—
19	75 – 90	90 – 100	100	—	—	—
16	65 – 83	78 – 92	90 – 100	100	—	—
13.2	57 – 76	62 – 80	76 – 92	90 – 100	100	—
9.5	45 – 65	50 – 72	60 – 80	68 – 85	90 – 100	100
4.75	24 – 52	26 – 56	34 – 62	38 – 68	45 – 75	90 – 100
2.36	16 – 42	16 – 44	20 – 48	24 – 50	30 – 58	55 – 75
1.18	12 – 33	12 – 33	13 – 36	15 – 38	20 – 44	35 – 55
0.6	8 – 24	8 – 24	9 – 26	10 – 28	13 – 32	20 – 40
0.3	5 – 17	5 – 17	7 – 18	7 – 20	9 – 23	12 – 28
0.15	4 – 13	4 – 13	5 – 14	5 – 15	6 – 16	7 – 18
0.075	3 – 7	3 – 7	4 – 8	4 – 8	4 – 8	5 – 10

4) Specimen preparation and specific gravity tests

The fourth step is to prepare specimens and test the specific gravities to obtain the four volumetric parameter characteristics, including, G_{mb}, VTM, VMA, and VFA, to determine the asphalt content.

(1) Specimen preparation

Specimen preparation includes preheating the aggregate in a bowl, adding hot asphalt, and mixing until the aggregate is well-coated. Then, we place a specific amount of mixed materials in the Marshall mold and use the Marshall hammer to compact 50 or 75 times on each side to make the specimen (Figure 7.34). Usually, we prepare specimens for five asphalt contents, and for each content, we make five specimens to ensure the optimum asphalt content will be covered in this range and to reduce the variation of the results.

7 Asphalt Mixtures

Figure 7.34　Preparation of the Marshall specimen

(2) Theoretical maximum specific gravity test

The first specific gravity that needs to test is the theoretical maximum specific gravity G_{mm}. It is only necessary to test the G_{mm} for one asphalt content, calculate the G_{se} with the obtained G_{mm}, and then calculate the G_{mm} for the other four groups.

(3) Bulk specific gravity test

After obtaining the G_{mm}, the bulk specific gravity of mixed aggregate G_{mb} can be calculated based on the proportions and the bulk specific gravity of each type of aggregates. It is the first volumetric characteristic for determination of asphalt content.

(4) Calculate volumetric characteristics

The three volumetric characteristics, *VTM*, *VMA*, and *VFA* can be calculated based on previously measured and calculated variables.

5) Marshall stability test

After preparing the specimens using the Marshall hammer, we firstly soak the specimens in the 60 ℃ water bath for 2 hours and then conduct the test to obtain the Marshall stability and the flow value.

6) Determination of optimum asphalt content

For the 5 different asphalt content, we can obtain the volumetric properties of the mixes including the bulk specific gravity G_{mb}, *VTM*, *VMA*, and *VFA*, as well as the mechanical properties of the mixes including Marshall stability and the flow value. The optimum asphalt content (OAC) is determined by checking the volumetric and mechanical properties of designed asphalt mixes at different asphalt content against the requirements of the specification. An example is presented below to go through the whole procedure. Firstly, we summarize the design requirements in Table 7.5 and draw the charts of the volumetric and mechanical characteristics compared with the asphalt content of diagrams in Figure 7.35.

Civil Engineering Materials

Table 7.5　Requirements for determining the optimum asphalt content

G_{mb}	VTM(%)	VMA(%)	VFA(%)	Marshall stability (kN)	Flow value (mm)
The higher the better	3–5	≥12	55–70	≥8	2–4

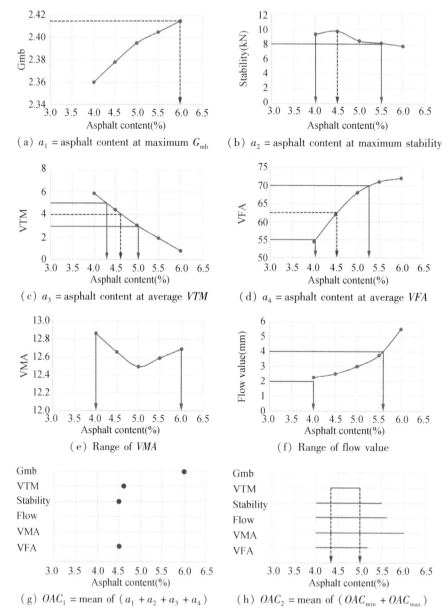

(a) a_1 = asphalt content at maximum G_{mb}　　(b) a_2 = asphalt content at maximum stability

(c) a_3 = asphalt content at average VTM　　(d) a_4 = asphalt content at average VFA

(e) Range of VMA　　(f) Range of flow value

(g) OAC_1 = mean of $(a_1 + a_2 + a_3 + a_4)$　　(h) OAC_2 = mean of $(OAC_{min} + OAC_{max})$

Figure 7.35　Determination of the optimum asphalt content in Marshall mix design

(1) The OAC_1 is calculated based on maximum density, maximum Marshall stability, an average of design VTM, and an average of design VFA.

- As shown in Figure 7.35(a), the asphalt content at the maximum G_{mb} is 6%.
- As shown in Figure 7.35(b), the stability requirement is higher than 8 kN. The corresponding range of asphalt content is 4% to 5.5%, and the asphalt content at the maximum stability is 4.5%.
- As shown in Figure 7.35(c), the VTM should be within 3% and 5%. The corresponding range of asphalt content is 4.3% to 5% and the asphalt content at the average VTM is 4.6%.
- As shown in Figure 7.35(d), the VFA should be within 55% and 70%, and the corresponding range of asphalt content is 4% to 5.3%, and the asphalt content at the average VFA is 4.5%.
- As shown in Figure 7.35(g), OAC_1 is the average of the four above content and equals 4.9%.

$$OAC_1 = \frac{a_1 + a_2 + a_3 + a_4}{4} \tag{7.32}$$

(2) The OAC_2 is the mid-value of the range that satisfies all requirements.

- As shown in Figure 7.35(e), the VMA should be higher than 12%, and all asphalt content meets this requirement. The corresponding range of asphalt content is from 4% to 6%.
- As shown in Figure 7.35(f), the flow value should be within 2 mm and 4 mm. The corresponding range of asphalt content is from 4% to 5.6%.
- As shown in Figure 7.35(h), the range that meets the requirement of all characteristics is from 4.3% to 5%, and OAC_2 is 4.7%.

$$OAC_2 = \frac{OAC_{min} + OAC_{max}}{2} \tag{7.33}$$

(3) The OAC is the average of OAC_1 and OAC_2, which is 4.8% in this example.

$$OAC = \frac{OAC_1 + OAC_2}{2} \tag{7.34}$$

7) Performance evaluation

The last step is to check the performance of the designed lab mix formula. The most frequently evaluated performance tests include moisture susceptibility, and high- and low-temperature stability. High-temperature stability is to test permanent deformation under specified times of wheel loads at high temperatures. Low-temperature stability is to test the cracking resistance of the mixture at low temperatures. The indirect tensile strength test and semi-circle bending test are frequently adopted. Moisture susceptibility is to test the strength retained after moisture conditioning, such as immersed Marshall stability test or freeze-thaw stability test. If the mix can not satisfy the performance check, we have to adjust the asphalt content or aggregate gradation to adjust the mix proportion to obtain the lab formula.

7.6 Superpave Mix Design

7.6.1 Background

Superpave means a superior performing asphalt pavement system. The Superpave mix design method was developed in the Strategic Highway Research Program (SHRP), aiming to tie asphalt and the aggregate selection into the mix design process, and considering traffic and climate as well. According to specification AASHTO R35-17, the Superpave mix design process consists of five steps, including the selection of aggregates, selection of binders, determination of the design aggregate structure, determination of the design binder content at 4% *VTM* and evaluation of moisture susceptibility.

In the Superpave mix design, specimens are prepared with the Superpave gyratory compactor. As shown in Figure 7.36, the mixture in the mold is placed in the compaction machine at an angle to the applied force. It uses a gyration angle of 1.16 degrees and a constant vertical pressure of 600 kPa. As the force is applied, the mold is gyrated, creating a shearing action in the mixture. Unlike the Marshall compactor that uses impact compaction which may crush aggregates, the gyratory compactor compresses the mixture slowly, simulating the actual compaction during construction, and therefore can produce better compaction quality for the specimen. The Superpave gyratory compactor produces specimens 150 mm in diameter and 95 mm to 115 mm in height, allowing the use of aggregates with a maximum size of more than 37.5 mm, while specimens prepared with the Marshall compactor are typically 101.6 mm in diameter and 63.5 mm in height, which limits the maximum aggregate size to 25 mm.

7 Asphalt Mixtures

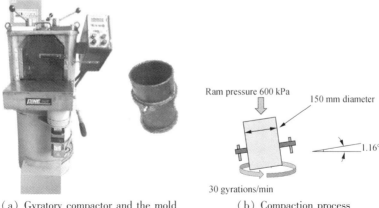

(a) Gyratory compactor and the mold (b) Compaction process

Figure 7.36　Superpave gyratory compactor

7.6.2　Procedures

1) Selection of aggregates

Aggregate for Superpave mixes must meet both source and consensus requirements. The source requirements include Los Angeles abrasion, soundness, and deleterious materials. The four tables show the consensus requirement of aggregate properties, including coarse aggregate angularity (Table 7.6), fine aggregate angularity (Table 7.7), flat and elongated particles (Table 7.8), and sand equivalency or clay content (Table 7.9). In Table 7.6, the first number is a minimum requirement for one or more fractured faces and the second number is a minimum requirement for two or more fractured faces. The consensus requirements are part of the national Superpave specification and are on the blend of aggregates.

Table 7.6　Requirement of minimum coarse aggregate angularity

20-year traffic (10^6 ESALs)	Depth from Surface	
	≤100 mm	>100 mm
<0.3	55/—*	—/—
0.3 – 3	75/—	50/—
3 – 10	85/80	60/—
10 – 30	95/90	80/75
≥30	100/100	100/100

Table 7.7　Requirements of minimum fine aggregate angularity

20-year traffic (10^6 ESALs)	Depth from surface	
	≤ 100 mm	> 100 mm
<0.3	—	—
0.3 – 3	40	40
3 – 10	40	40
10 – 30	45	45
≥30	45	45

Table 7.8　Requirements of flat and elongated particles

20-year traffic (10^6 ESALs)	Maximum percentage of particles with thickness >5
<0.3	—
0.3 – 3	10
3 – 10	10
10 – 30	10
≥30	10

Table 7.9　Requirements of sand equivalency or clay content

20-year traffic (10^6 ESALs)	Minimum sand equivalency (%)
<0.3	40
0.3 – 3	40
3 – 10	45
10 – 30	45
≥30	50

Aggregates used in asphalt concrete must be well graded. The 0.45 power chart is recommended for Superpave mix design (Figure 7.37). The gradation curve must go between control points. The restricted zone requirement was initially adopted in Superpave to help reduce the rutting potential. However, according to many asphalt paving technologists, compliance with the restricted zone criteria may not be desirable or necessary to produce mixtures that give a good performance, and therefore it is not used anymore.

7 Asphalt Mixtures

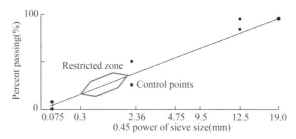

Figure 7.37 Control points and the restricted zone in Superpave gradations

2) Selection of binders based on viscosity-temperature

The binder is selected based on the maximum and minimum pavement temperatures, which are the high 7-day average temperature and the low 1-day lowest temperature respectively, as discussed in the asphalt PG grading. In addition to the specification tests, specific gravity is needed for the volumetric analysis. The viscosity-temperature relationship is needed to determine the required mixing and compaction temperatures. The proper viscosity for mixing is 0.17 Pa · s, and that for compaction is 0.28 Pa · s. After selecting the original binder grade based on temperature, Superpave suggests adjusting the binder grade based on the traffic volume. Table 7.10 suggests an increase of one grade for slow transient loads and high traffic volume roads and two grades for stationary loads.

Table 7.10 The binder grade based on traffic conditions

Original grades	Grades for slow transient loads (+1 grade)	Grades for stationary loads (+2 grades)	20-year traffic (10^6 ESALs) (+1 grade)
PG 58 - 22	PG 64 - 22	PG 70 - 22	PG 64 - 22
PG 70 - 22	PG 76 - 22	PG 82 - 22	PG 76 - 22

3) Aggregate structure

When checking aggregate structure, trial specimens are prepared with three different aggregate gradations. As shown in Table 7.11, the number of gyrations for compaction is determined by the number of predicted ESALs. The Superpave method recognizes three critical stages of compaction, including initial, design, and maximum.

(1) $N_{initial}$ is the number of gyrations used as a measure of mixture compactibility during construction. Mixtures with low air voids at $N_{initial}$ are tender during construction and unstable when subjected to traffic. For example, a mixture designed for greater than or equal to 3 million ESALs should have at least 11 % air voids at $N_{initial}$.

(2) N_{design} is the design number of gyrations required to produce a specimen with the same density as that expected in the field after the indicated amount of traffic. 4% air voids at N_{design} is desired in the mix design.

(3) N_{max} is the number of gyrations required to produce a laboratory density that should never be exceeded in the field. If the air voids at N_{max} are too low, the field mixture may be compacted too much under traffic, resulting in excessively low air voids and potential rutting. The air void content at N_{max} should never be below 2% air voids. Typically, specimens are compacted to N_{design} to establish the optimum asphalt binder content and then additional specimens are compacted to N_{max} as a check.

Table 7.12 summarizes the requirements for Superpave mix design criteria. The percentage of theoretical maximum specific gravity at different gyrations for different traffic levels is 100 minus air voids or VTM. For example, for 3 – 10 million design ESALs, the VTM at $N_{initial}$ should be no less than 11%, the VTM at N_{design} should be 4%, and the VTM at N_{max} should be no less than 2%. The VFA should be between 65% and 75%. In addition, the ratio of dust to effective asphalt should be between 0.6 and 1.2. And the tensile strength ratio from the freeze-thaw moisture susceptibility test should be no less than 80%. The VMA should be no less than a value for different nominal maximum sizes.

Table 7.11 Number of gyrations for different traffic levels

Number of gyrations	20-year traffic (10^6 ESALs)				
	<0.3	0.3 – 3	3 – 10	10 – 30	≥30
$N_{initial}$	6	7	8(7)	8	9
N_{design}	50	75	100 (75)	100	125
N_{max}	75	115	160 (115)	160	205

Table 7.12 Superpave mix design criteria

20-year traffic (10^6 ESALs)		<0.3	0.3 – 3	3 – 10	10 – 30	≥30
Percentage of theoretical maximum specific gravity	$N_{initial}$	91.5	90.5	89	89	89
	N_{design}	96	96	96	96	96
	N_{max}	98	98	98	98	98
VFA		70 – 80	65 – 78	65 – 75	65 – 75	65 – 75
Design air voids		0.04				
Dust to effective asphalt		0.6 – 1.2				
Tensile strength ratio		80% min				
Nominal maximum size (mm)		37.5	25	19	12.5	9.5	4.75
Minimum VMA (%)		11	12	13	14	15	16

7 Asphalt Mixtures

4) Binder content

The design binder content is obtained by preparing eight specimens, with two replicates at each of four binder content: estimated optimum binder content, 0.5% less than the optimum, 0.5% more than the optimum, and 1% more than the optimum. The volumetric properties are calculated and plots are prepared of each volumetric parameter versus the binder content (Figure 7.38). The binder content that corresponds to 4% VTM is determined. Other plots are used to check the volumetric parameters at the selected binder content.

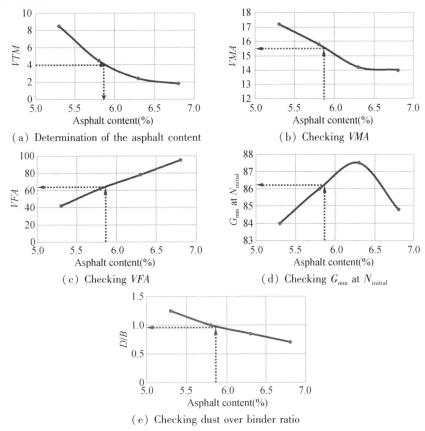

Figure 7.38 Determination of the asphalt content in Superpave mix design

5) Moisture susceptibility

If all the properties meet the criteria, six specimens are prepared at the design binder content and 7% air voids for the moisture susceptibility check. Three specimens are conditioned by vacuum saturation, then freezing and thawing; three other specimens are not conditioned. The tensile strength of each specimen is measured and the *TSR* is determined. The minimum Superpave criterion for *TSR* is 80%.

7.7 Typical Mixtures

7.7.1 Dense-Graded Mixtures

Dense-graded mixtures were popular because they required relatively low asphalt binder content, and therefore low costs. Currently, dense-graded mixtures have a void content of 3%–7% and an asphalt content of 4.5%–6%. They have a good aggregate interlock and relatively low permeability. Dense-graded mixtures are used for the base, binder, and surface courses. The coarser mixtures (AC-25, AC-20) are used for the base and binder courses, and the finer mixtures (AC-16, AC-13, and AC-10) are used for top courses. The procedure for designing dense-graded HMA mixtures suggests that a range of gradations be evaluated and the gradation which is most effective in meeting the given mixture specifications should be selected. The specification JTG F40—2004 recommends the gradation of dense-graded mixtures and the minimum *VMA* as shown in Table 7.13 and Table 7.14, respectively.

Table 7.13　The recommended gradation of dense-graded mixtures

Sieve size (mm)	Passing (%)					
	AC-25	AC-20	AC-16	AC-13	AC-10	AC-5
31.5	100	—	—	—	—	—
26.5	90 – 100	100	—	—	—	—
19	75 – 90	90 – 100	100	—	—	—
16	65 – 83	78 – 92	90 – 100	100	—	—
13.2	57 – 76	62 – 80	76 – 92	90 – 100	100	—
9.5	45 – 65	50 – 72	60 – 80	68 – 85	90 – 100	100
4.75	24 – 52	26 – 56	34 – 62	38 – 68	45 – 75	90 – 100
2.36	16 – 42	16 – 44	20 – 48	24 – 50	30 – 58	55 – 75
1.18	12 – 33	12 – 33	13 – 36	15 – 38	20 – 44	35 – 55
0.6	8 – 24	8 – 24	9 – 26	10 – 28	13 – 32	20 – 40
0.3	5 – 17	5 – 17	7 – 18	7 – 20	9 – 23	12 – 28
0.15	4 – 13	4 – 13	5 – 14	5 – 15	6 – 16	7 – 18
0.075	3 – 7	3 – 7	4 – 8	4 – 8	4 – 8	5 – 10

Table 7.14 The minimum requirements of VMA for dense-graded mixture

Mixture types		AC					ATB		
NMAS (mm)		4.75	9.5	13.2	16	19	26.5	26.5	31.5	37.5
Target void content (%)	2	15.0	13.0	12.0	11.5	11.0	10.0	—		
	3	16.0	14.0	13.0	12.5	12.0	11.0			
	4	17.0	15.0	14.0	13.5	13.0	12.0	12.0	11.5	11.0
	5	18.0	16.0	15.0	14.5	14.0	13.0	13.0	12.5	12.0
	6	19.0	17.0	16.0	15.5	15.0	14.0	14.0	13.5	13.0

7.7.2 Gap-Graded Mixtures

The SMA was developed in Germany to resist the wear of studded tires in winter and rutting in summer. SMA is based on the interlock of coarse aggregates leaving a relatively high void content between aggregates particles. This void space is partly filled with a binder-rich mastic mortar, consisting of fine aggregates, fillers, and modified asphalt. To ensure the stone-on-stone contact, two requirements should be met: (1) the minimum *VMA* of the compacted mixture should be not less than 17%; (2) the voids in the coarse aggregate of the mix (VCA_{mix}) should be no more than the voids of coarse aggregate dry-rodded (VCA_{dry}). The designing of an SMA mixture involves adjusting the aggregate gradation to accommodate the required asphalt and void content while the traditional asphalt mixture design is mainly to adjust the asphalt content for a selected aggregate gradation. SMA is widely used as the top layer for heavy traffic in China. The specification JTG F40—2004 recommends the gradation of SMA as shown in Table 7.15.

Table 7.15 The recommended gradation of SMA

Sieve size (mm)	Passing (%)			
	SMA-20	SMA-16	SMA-13	SMA-10
26.5	100	—	—	—
19	90 – 100	100	—	—
16	72 – 92	90 – 100	100	—
13.2	62 – 82	65 – 85	90 – 100	100
9.5	40 – 55	45 – 65	50 – 75	90 – 100
4.75	18 – 30	20 – 32	20 – 34	28 – 60

Continued

Sieve size (mm)	Passing (%)			
	SMA-20	SMA-16	SMA-13	SMA-10
2.36	13 – 22	15 – 24	15 – 26	20 – 32
1.18	12 – 20	14 – 22	14 – 24	14 – 26
0.6	10 – 16	12 – 18	12 – 20	12 – 22
0.3	9 – 14	10 – 15	10 – 16	10 – 18
0.15	8 – 13	9 – 14	9 – 15	9 – 16
0.075	8 – 12	8 – 12	8 – 12	8 – 13

7.7.3 Open-Graded Mixtures

The open-graded mixture is known as the porous asphalt (PA) mixture and has a high interconnected air-voids content after compaction, typically between 18%–25%. In the US, it is usually called the "open-graded friction course (OGFC)" and is mainly used in the surface layer. PA reduces splashing, spray, and glare, and improves skid resistance and visibility of pavement markings. The interconnected voids also allow the movement of air, reducing noise. When porous asphalt is used as the surface course, the binder course is impermeable to protect the remainder of the pavement and its foundation from the damaging effect of the water. There are several special applications of PA:

(1) Asphalt treated permeable base (ATPB) is an open-graded mixture used for the base and binder courses. It is free-draining, has high stone-on-stone contact for stability, and is resistant to fatigue and reflective cracking. When used as a base course, a moisture seal should be provided under the open-graded mixture to prevent water from entering and weakening the subgrade materials.

(2) The sustainable drainage system (SuDS) uses PA to alleviate the problems of excess water in urban areas with extensive areas of impermeable surfacing by storing and/or slowing down the flow of surface water. SuDS is generally used for parking lots and lightly traffic areas.

(3) The cement grouted bituminous macadam is a composite material in which the voids in a PA mixture are filled with a cementitious grout. The highly permeable material is converted into a highly impermeable material. It is used for heavily loaded areas and locally on PA surface courses in areas of high stress such as roundabouts.

PA requires single-sized coarse aggregates to ensure high interconnected air voids and good inter-locking. The aggregate gradation mainly consists of coarse aggregates: about 80% retained on the 2.36 mm sieve, as shown in Table 7.16. Generally, the amount of asphalt in PA is greater and either fiber or modified asphalt is needed. The cost of PA is higher than that of AC or SMA.

Table 7.16 The recommended gradation of porous asphalt mixtures

Sieve size (mm)	Passing (%)					
	Mixes for surface layer			Mixes for base layer		
	OGFC-16	OGFC-13	OGFC-10	ATPB-40	ATPB-30	ATPB-25
53	—	—	—	100	—	—
37.5	—	—	—	70 – 100	100	—
31.5	—	—	—	65 – 90	80 – 100	100
26.5	—	—	—	55 – 85	70 – 95	80 – 100
19	100	—	—	43 – 75	53 – 85	60 – 100
16	90 – 100	100	—	32 – 70	36 – 80	45 – 90
13.2	70 – 90	90 – 100	100	20 – 65	26 – 75	30 – 82
9.5	45 – 70	60 – 80	90 – 100	12 – 50	14 – 60	16 – 70
4.75	12 – 30	12 – 30	50 – 70	0 – 3	0 – 3	0 – 3
2.36	10 – 22	10 – 22	10 – 22	0 – 3	0 – 3	0 – 3
1.18	6 – 18	6 – 18	6 – 18	0 – 3	0 – 3	0 – 3
0.6	4 – 15	4 – 15	4 – 15	0 – 3	0 – 3	0 – 3
0.3	3 – 12	3 – 12	3 – 12	0 – 3	0 – 3	0 – 3
0.15	3 – 8	3 – 8	3 – 8	0 – 3	0 – 3	0 – 3
0.075	2 – 6	2 – 6	2 – 6	0 – 3	0 – 3	0 – 3

7.7.4 Poured Asphalt Mixtures

Poured asphalt mixtures are placed using their weight, similar to the Portland cement concrete. There are two main products: mastic asphalt (MA) and gussasphalt. The MA is mainly used in Britain, France, and the Mediterranean. It is a mortar-based asphalt mixture with a maximum fine aggregate size of 2 mm and 10/14 mm coarse aggregate particles embedded as "plums" and 8/14 mm or 14/20 mm pre-coated chippings when used as the surface course. With the very high binder content, MA is the most impermeable of the asphalt types and, with the stiff binder, reasonably deformation-resistant. MA is mainly used in concrete bridge decks, steel bridge decks, and tunnels where there is a need for a water-resistant material that can be laid thinly and with a smooth finish.

The gussasphalt is mainly used in Germany and Northern Europe. Gussasphalt was developed in Germany and can be used on a wide range of roads in addition to bridges and tunnels, while MA is limited to applications on structures. The production of gussasphalt involves pre-blending the asphalt and aggregates for two minutes at around 210 – 220 ℃. It needs modified plant and paving machines trucks as shown in Figure 7.39.

Figure 7.39 The guss-cooker and paver for gussasphalt

7.8 Construction

The production and construction of the asphalt mixture pavement can be described as a four-step process: manufacture, transportation, paving, and compaction.

7.8.1 Manufacture

The manufacture of the asphalt mixture includes drying aggregates, blending them in proportions based on the formula, adding asphalt, and mixing. There are two types of asphalt plants: the batch plant and the drum plant (Figure 7.40). The batch plant produces the asphalt mixture in batches while the drum plant produces the asphalt mixture in a continuous process. The major difference is that the batch plant heats aggregates and uses hot sieves to separate aggregates while the drum plant heats aggregates after cold sieving aggregates and the moisture content is adjusted.

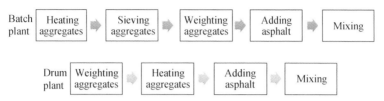

Figure 7.40 Mixing process of the batch plant and drum plant

7.8.2 Transportation

In a batch plant, aggregates are transported from the stockpiles to the cold feed bins and then transported to the dryer which is a large rotating cylinder with flights that promote tumbling of aggregates through heated air. The hot and dry aggregates tumble from the dryer to the hot elevator which lifts aggregates to the hot sieves. The hot sieves separate aggregates and charge the hot bins. The required amount of aggregates is weighed out from each hot bin into the mixer. Aggregates are briefly mixed, then the hot asphalt is added, and aggregates and asphalt are thoroughly mixed. The HMA is then discharged into a silo or directly into trucks. In a drum plant, a single drum is used for both drying the aggregates and mixing the asphalt cement. Weight scales are used on the cold feed conveyor belts to monitor the amount of aggregates from each stockpile and the weights are adjusted for the moisture in aggregates. Aggregates are dried in the first two-thirds and asphalt is added in the last one-third, creating a continuous mixing process. The HMA is then transported using a hot elevator into a silo and then loaded into trucks.

7.8.3 Paving

Asphalt mixtures will lose heat in the process of transportation, resulting in dispersion of the mixture. Therefore, the truck with covers in the transportation of asphalt mixtures is used to prevent heat loss. Traditionally, trucks will directly load materials into the paver's hopper. This has several potential problems such as the truck bumping the paver, non-uniform delivery of materials, and the potential for temperature segregation of the mix. In response to these problems, the industry has developed material transfer devices. The use of a material transfer device is optional, but it is becoming common for paving sections where smoothness is critical such as on interstates and airfield pavements. The next step is to put the asphalt mixture on the road. Asphalt pavers are designed to produce a smooth surface when operated properly. The keys to proper operation are a constant flow of the material through the paver and maintaining constant paver speed.

7.8.4 Compaction

It is necessary to control how much compaction effort is required. The optimum air voids during compaction are specified based on the fact that traffic will continue to densify the pavement. If the air voids drop below 2% of the mix during compaction, the mix may become unstable, resulting in bleeding, corrugations, flushing, early-age rutting, or low skid resistance problems. But more than 8% of air voids are also detrimental to the performance of the mix, as the air voids become interconnected which creates problems with the infiltration of both air and water into the asphalt mat. Compaction is a three-stage process:

(1) Breakdown rolling is one or two passes of the roller to set the mat, and most of the density is achieved during the breakdown rolling.

(2) Intermediate rolling is when further compaction is needed to achieve the desired density.

(3) Finish rolling is to remove roller marks left on the surface by the previous passes of the roller.

Due to the change in the viscosity of the asphalt with temperature, compaction must be achieved while the asphalt viscosity is low enough so that the asphalt acts as a lubricant to allow arranging aggregates into a dense configuration. Moreover, at lower temperatures the viscosity of the asphalt inhibits compaction.

Several types of rollers are used for asphalt paving: static steel wheel, vibratory steel wheel, or rubber wheel. Vibratory rollers are becoming the dominant type of machine as they can be used for all stages. With the vibrators active, high forces are applied to the pavement to achieve density. The vibrators can be turned off for the finish rolling.

7.8.5 Recycling

Recycling pavement materials have a long history. However, recycling became more important in the mid-1970s, after the oil embargo, due to the increase in asphalt prices. There are two types of recycling, plant recycling, and in-place recycling. Plant recycling is performed by milling the old pavement and sending the reclaimed asphalt pavement (RAP) to a central asphalt concrete plant, where it is mixed with the virgin aggregates and asphalt binders in the asphalt plant. In-place recycling finishes milling, mixing, and paving on site, therefore, includes all the equipment on site. Moreover, it can be further classified into cold in-place recycling and hot in-place recycling depending on the mixing technology.

Questions

1. What is an asphalt mixture? What is the difference between the asphalt concrete mixture and the asphalt macadam mixture?

2. How to classify asphalt mixtures? What is the significance of these classifications in production and application?

3. How to classify asphalt mixtures by their structures? What are the advantages and disadvantages of different structural types?

4. Discuss the principles of strength formation of the asphalt mixture, and analyze them from the internal material composition parameters and external influencing factors.

5. How to evaluate the high-temperature stabilities of the asphalt mixture? Briefly describe the test methods and the indices.

6. For an asphalt mixture with an asphalt content of 5%, a bulk specific gravity of 2.382, a theoretical maximum specific gravity of 2.480, a void content of 4%, and an aggregate bulk specific density of 2.705. What is the VFA?

7. What are the causes of thermal cracks in the asphalt mixture and how to evaluate the anti-cracking properties of the asphalt mixture?

8. Discuss the Marshall design methods for HMA. How to determine the composition of mineral mixtures and the optimum amount of asphalt?

9. Briefly introduce how to determine the optimum asphalt content of HMA using the Superpave method.

10. Discuss the main properties of the asphalt mixture and the current methods to evaluate those properties.

8 Steel

The metal material is a general term for an alloy consisting of one or more metallic elements and some nonmetallic elements. Metal materials exhibit high strength, large elastic modulus, dense structure, and capability of casting, welding, assembling, and mechanical construction. Therefore, metal is an important building material. Metal materials are generally classified into ferrous metals and non-ferrous metals. Ferrous metals contain iron as their main constituent or base metal and are the most used metals in civil engineering. Non-ferrous metals are other metals except ferrous metals, such as aluminum, lead, zinc, copper, tin, and their alloys. The aluminum alloy is an important lightweight structural material. This chapter introduces the production, mechanical properties, hot and cold treatment, structural steel, and the corrosion of steel.

8.1 Production

The use of iron dates back to about 3500 years ago when primitive furnaces were used to heat the ore in a charcoal fire. Ferrous metals were produced on a relatively small scale until the blast furnace was developed in the 1700s. Iron products were widely used in the latter half of the 18th century and the early part of the 19th century. Steel production started in the mid-1800s when the Bessemer converter was invented. In the second half of the 19th century, steel technology advanced rapidly due to the development of the basic oxygen furnace and continuous casting methods. More recently, computer-controlled manufacturing has increased efficiency and reduced the cost of steel production. China's annual steel production has ranked first in the world since 1996, and currently produces and uses 50 % of the world's steel. The overall process of steel production consists of the following three phases:

(1) Iron making is to reduce iron ore to pig iron. Iron contains more carbon and other impurities. The carbon content of iron is more than 2.06%, while that of steel is less than 2.06%.

(2) Steel making is to refine pig iron and scrap steel from the recycling of steel.

(3) Steel rolling is to cast and roll steel into products including the bar, wire, pipe, sheet, etc.

8.1.1 Iron Production

Iron, which makes up 4.8% of the Earth's crust, is usually found in iron ore as a chemical compound. The iron ore includes hematite ore (Fe), magnetite (Fe_3O_4), siderite ($FeCO_3$), limonite ($Fe_2O_3 \cdot 2Fe(OH)_3$), and pyrite (FeS_2). Coal and limestone are also used to produce pig iron. The coal, after being transformed into coke, supplies carbon to reduce the amount of oxygen in the ore. Limestone is used to help remove impurities. The concentration of iron in the ore is increased by crushing and soaking the ore. The processed ore contains about 65% iron. Reduction of the ore to pig iron is accomplished in a blast furnace (Figure 8.1). Oxygen in the ore reacts with carbon to form carbon oxides. A flux is used to help remove impurities. The molten iron is at the bottom of the furnace. The impurities and slag float on top of the molten iron.

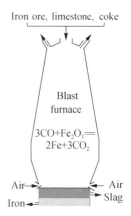

Figure 8.1 Iron production in a blast furnace

8.1.2 Steel Production

Steel is produced by removing excess carbon and other impurities in iron. Refining pig iron to steel needs an oxygen furnace or arc furnace (Figure 8.2). The basic oxygen furnaces remove excess carbon by reacting the carbon with oxygen. The steel production process is continued until all impurities are removed and the desired carbon content is achieved. Electric furnaces use an electric arc between carbon electrodes to melt and refine the steel. It can heat charged pig iron by an electric arc to 1800 ℃.

Civil Engineering Materials

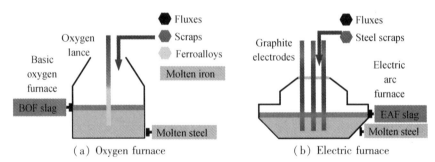

Figure 8.2 Steel production

8.2 Classification

Steel can be classified into different types based on its composition and physical properties.

8.2.1 Deoxygenation

During steel production, oxygen may be dissolved in the liquid metal. As steel solidifies, oxygen can react with carbon to form carbon monoxide bubbles that are trapped in the steel and cause fractures. Deoxidizing agents, such as aluminum, ferrosilicon, and manganese, can be used to eliminate the formation of carbon monoxide bubbles.

1) Rimmed steel

Rimmed steel does not have deoxidizing agents added to it during casting which causes carbon monoxide to evolve rapidly from the ingot. Rimmed steel has uneven composition and poor compactness, which influences the quality of steel. However, its cost is low and can be widely used in general building structures.

2) Killed steel

Killed steel is steel that has been completely deoxidized by the addition of an agent before casting so that there is practically no evolution of gas during solidification. Killed steel has better quality than rimmed steel, but costs more. It is mostly used in parts or joints bearing impact loads.

3) Semi-killed steel

Semi-killed steel is mostly deoxidized steel, but the carbon monoxide leaves blow-hole type porosity distributed throughout the ingot. Semi-killed steel is commonly used for structural steel with a carbon content between 0.15% and 0.25%, because it is rolled, which closes the porosity.

8.2.2 Chemical Composition

Carbon steel is also known as plain steel. Carbon steel is an iron-carbon alloy with a carbon content of less than 1.35%, a silicon content of less than 0.4%, and a small amount of sulfur and phosphorus impurities. Carbon plays a major role in the performance of steel, while other elements, such as silicon, sulfur, and phosphorus, do not play decisive roles due to their low contents. According to the amount of carbon, carbon steel can be classified into low, medium, and high carbon steel.

(1) Low carbon steel has a carbon content of less than 0.25%. It is soft and easy to work with, but cannot be quenched and annealed. It is the main steel used in civil engineering.

(2) Medium carbon steel has a carbon content of 0.25%–0.6%. It is hard, can be quenched and annealed, and is mostly used as mechanical parts.

(3) High carbon steel has a carbon content of higher than 0.6%. It is very hard, can be quenched and annealed, and is mainly used for tools.

According to the content of phosphorus, sulfur, and other impurities in the steel, carbon steel can also be classified into:

(1) Plain carbon steel with phosphorus and sulfur content of less than 0.045% and 0.050%, respectively;

(2) Quality carbon steel with phosphorus and sulfur content of both less than 0.035%;

(3) High-quality carbon steel with phosphorus and sulfur content of both less than 0.025%.

Alloy steel is to add different alloy metals to carbon steel to alter its characteristics. According to the content of alloying elements, it is classified into:

(1) Low alloy steel with a total alloy content of less than 3.5%;

(2) Medium alloy steel with a total alloy content of 3.5%–10%;

(3) High alloy steel with a total alloy content of higher than 10%.

8.2.3 Applications

According to different applications, steel can be classified into structural steel, tool steel, special steel, and functional steel.

1) Structural steel

Structural steel is a type of steel that is used as a construction material. According to chemical composition, structural steel is classified into structural carbon steel and structural alloy steel.

(1) Structural carbon steel includes plain and superior structural carbon steel.

① Plain structural carbon steel contains carbon of less than 0.38%. They are used to produce bars, square steel, flat steel, angle steel, slot steel, steel plates, etc.

② Superior structural carbon steel contains fewer impurities. It has better performance and is commonly used for mechanical parts, tools, springs, etc. It can be further classified into pressure processing steel and cutting processing steel.

(2) Structural alloy steel includes low structural alloy steel and structural alloy steel.

① Low structural alloy steel is the plain structural carbon steel with a few additional alloys. It has high strength, resilience, and weldability. It is widely used to produce rebars and steel plates.

② Structural alloy steel includes spring alloy steel, ball bearing steel, manganese steel, chrome steel, nickel steel, boron steel, etc. They are mainly used to build machines and equipment.

2) Tool steel

Tool steel refers to a variety of carbon steel and alloy steel that are particularly well-suited to be made into tools. It has distinctive hardness, resistance to abrasion and deformation, and the ability to hold a cutting edge at elevated temperatures. Tool steel can be classified into carbon tool steel, alloy tool steel, and high-speed tool steel. It is used in various fields, such as cutting, molding, and measuring tools.

(1) Carbon tool steel typically has a carbon content of 0.65%–1.35%. It is further classified into superior and advanced carbon steel according to the content of sulfur and phosphorus. Carbon tool steel can be used to produce steel brazing and hollow steel brazing.

(2) Alloy tool steel contains a higher content of carbon. It exhibits high hardness,

abrasion resistance, and small heat deformation at high temperatures. Alloy tool steel is classified into measuring and cutting alloy tool steel, abrasion-resistant alloy tool steel, cold-rolled mold steel, and hot-rolled mold steel.

(3) High-speed tool steel has a high alloy content and has superior hardness and abrasion resistance to regular tool steel. It is mainly used for drilling and cutting tools such as the power-saw blades and drill bits.

3) Special steel

Special steel is mostly high alloy steel, including stainless steel, heat-resistant steel, abrasion-resistant steel, and electrical silicon steel.

4) Functional steel

Functional steel is classified into carbon steel and alloy steel, including rebar, bridge steel, rail steel, boiler steel, mining steel, and marine steel.

(1) Steel rebar includes mainly low alloy steel and ribbed steel for reinforced concrete.

(2) Bridge steel typically uses killed steel to obtain strength and high-impact resilience. It is classified into carbon steel and plain alloy steel.

(3) Rail steel is classified into heavy and light rail steel. Heavy rail steel uses killed steel, while light rail steel uses killed or semi-killed steel.

8.2.4 Steels in Civil Engineering

1) Carbon structural steel

Carbon structural steel is widely used in civil engineering and is suitable for producing various types of steel bars, wires, and sections. It is also directly used in the hot rolled process. Steel products are typically labeled with Chinese phonetic letters, element symbols, and numbers, indicating the grade, application, properties, and manufacture of the steel. In China, the grade for carbon steel includes four parts:

(1) Alphabet "Q" representing yield strength.

(2) Yield strength value in the unit of MPa, including five levels, ranging from 195 to 275 MPa.

(3) The quality grades of A, B, C, and D. Grade A steel has the highest content of impurities. Grade D steel has the lowest amount of impurities.

(4) Deoxidation methods of F, B, and Z. F means rimmed steel, B means semi-killed steel, and Z means killed steel or completely deoxidized steel.

For example, "Q235AF" means a quality grade A rimmed carbon steel with 235 MPa yield strength. According to specification GB/T 700—2006, carbon structural steel includes five types: Q195, Q215, Q235, Q255, and Q275.

2) Superior carbon structural steel

Compared with normal carbon structural steel, superior carbon structural steel contains fewer harmful impurities such as sulfur and phosphorous and non-metallic inclusions. It has higher plasticity and toughness, can be strengthened by heat treatment, and is mostly used in more important structures. In civil engineering, it is often used to produce steel wires, steel strands, high-strength bolts, prestressed anchors, etc. The grade for superior carbon structural steel includes three parts:

(1) Two digits of carbon content in the unit of 0.01%.

(2) Deoxidation method in which none means killed steel, and F means rimmed steel.

(3) Manganese (Mn) content, in which none means content lower than 0.8% while Mn means 0.7%–1.2%. With the increase of Mn content, the strength and hardness increase, while plasticity and toughness decrease.

For example, "50Mn" means quality carbon steel with 0.5% carbon and 0.7%–1.2% Mn. According to specification GB/T 699—2015, there are 28 types of superior carbon structural steel with different carbon content (0.05% to 0.75%).

3) Low alloy structural steel

Low alloy structural steel is high-strength steel formed by adding usually less than 3% alloy to low carbon steel. The main alloy elements include manganese (Mn), silicon (Si), vanadium (V), titanium (Ti), niobium (Nb), and other rare elements (RE). Low alloy structural steel is widely used in bridges, vehicles, and rebar production because of its high strength, wear resistance, and corrosion resistance. The grade for low alloy structural steel includes three parts:

(1) Alphabet "Q" representing yield strength in Chinese.

(2) Yield strength value in the unit of MPa.

(3) Quality grades, including A, B, C, and D. Grade A has the highest content of impurities such as sulfur and phosphorus, followed by B, C, and D.

For example, "Q295D" means quality grade D low alloy steel with 235 MPa yield strength. According to specification GB/T 1591—2018, there are eight types of low alloy structural steel, including Q355, Q390, Q420, Q460, Q500, Q550, Q620, and

Q690. The grade Q460E low alloy structural steel was used to build National Stadium (Bird's Nest) (Figure 8.3).

Figure 8.3 National Stadium (Bird's Nest)

4) Alloy structural steel

Due to the effect of alloy elements, the strength, toughness, and wear resistance of alloy steel are significantly improved. The alloy structural steel is widely used in high-strength bolts, gears, and other important structural components. Alloy structural steel usually requires heat treatment to obtain good mechanical performance. The grade for alloy structural steel includes 3 parts.

(1) Two digits of carbon content in the unit of 0.01%.

(2) An element symbol representing the major alloy.

(3) A number representing alloy content in the unit of 1% when it is higher than 1.6%.

For example, "20SiMn2MoV" means 2% carbon content alloy structural steel including 2% manganese as well as silica and molybdenum. According to specification GB/T 3077—2015, there are 86 types of alloy structural steel, including 40Mn2, 20SiMn2MoV, 40Cr, etc.

5) Bridge structural steel

Bridge structural steel is the steel used specifically for building bridges. Steel bridge structures can be built quicker due to their modular design. In most instances, steel beams and other small elements are made off-site and then transported to the jobsite. The grade of steel bridge is the same as quality carbon steel but ended with "q" representing it is for bridge use. According to specification GB/T 714—2015, the eight types of bridge structural steel include Q345q, Q370q, Q420q, Q460q, Q500q, Q550q, Q620q, and Q690q.

8.3 Chemical Composition

The properties of steel are mainly determined by its chemical composition. Steel is an alloy made up of iron with a little carbon and may contain other elements such as silicon, manganese, phosphorus, sulfur, oxygen, nitrogen, etc. Those elements have different influences on the properties of steel.

8.3.1 Influence of Elements

1) Carbon

Carbon is the most abundant element in steel except iron. As shown in Figure 8.4, with the increase of carbon content, the hardness and strength increase while the plasticity, toughness, and cold bending properties decrease. When the carbon content increases to about 0.8%, the strength reaches the maximum, but when the carbon content exceeds 0.8%, the strength decreases. Part of the carbon in the steel dissolves into iron as a

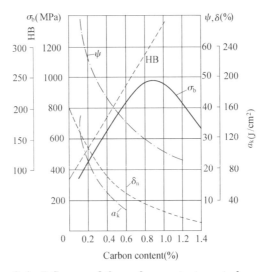

Figure 8.4　Influence of the carbon content on steel properties

(HB = Brinell hardness, σ_b = tensile strength, δ_n = elongation at fracture (%),
ψ = reduction of area at fracture (%), α_k = impact toughness)

solid solution, while the other part is combined with iron to form cementite. Carbon steel is a mixture of low hardness and high plasticity ferrite, and hard and brittle cementite. Therefore, the tensile strength and hardness of carbon steel increase linearly with the increase of the carbon content, while the plasticity of carbon steel decreases.

The influence of carbon on steel is not only related to the hardness and brittleness of cementite itself but also related to the distortion of the crystal lattice at the interface between cementite and ferrite. The grain interface between cementite and ferrite restains the slip of ferrite and causes micro cracks. Therefore, the increase of cementite improves the grain interface and the deformation resistance, while decreases the plasticity.

2) Silicon

Silicon is added to the steel as a deoxidizer to capture oxygen from FeO and form oxides to reduce the oxygen content in the steel. Silicate can be further formed by oxides of silicon and basic oxides in molten steel as shown in Equations (8.1) and (8.2). When the molten steel is poured, silicate impurities have no time to float into the slag and remain in the steel. It is easy to concentrate in the grain boundary during solidification, and the steel is easy to deform or fracture during pressure processing. The content of silicon in carbon steel is 0.1%–0.4%. The incorporation of silicon into ferrite increases the strength and hardness of steel significantly although the silicon content in carbon steel is very low. When the content of silicon increases to 1%–1.2% as an alloying element, the tensile strength of the steel can be increased by 15%–20%, but the plasticity, toughness, and weldability decrease.

$$SiO_2 + CaO \Longrightarrow CaSiO_3 \qquad (8.1)$$

$$SiO_2 + Fe_2O_3 + C \Longrightarrow CO \uparrow + Fe_2SiO_4 \qquad (8.2)$$

3) Manganese

Manganese is added to steel as a deoxidizer as shown in Equation (8.3). The content of manganese in carbon steel is 0.25%–0.8%. When the content of manganese is less than 0.8%, the effect of Mn on the properties of steel is not significant. When manganese is added as an alloy to a content of 0.8%–1.2% or higher, the mechanical properties of steel will be significantly improved. However, when the manganese content is greater than 1%, the corrosion resistance and weldability of steel decrease.

$$Mn + FeO \Longrightarrow MnO + Fe \qquad (8.3)$$

4) Phosphorus

When phosphorus is dissolved into ferrite, the strength, and hardness of steel increase, while the plasticity and toughness decrease significantly. When the content of phosphorus is higher than 0.3%, the steel becomes completely brittle and the impact toughness is close to zero. This phenomenon is called cold brittleness. Although the content of phosphorus is usually low, the phosphorus in the steel is easy to segregate during crystallization, causing the local area with a higher phosphorus content to become brittle. Therefore, the content of phosphorus is strictly controlled to be less than 0.045%. Phosphorus can improve the cutting performance of steel, and the cutting tool steel contains more phosphorus.

5) Sulfur

Sulfur can form ferrous sulfate. Ferrous sulfate and ferrite then can form eutectic at 985 ℃ and causing cracking above 1000 ℃ due to eutectic melting, which is called hot brittleness. The sulfur content is also strictly controlled to be less than 0.055%. The harmful influence of sulfur can be reduced by the addition of manganese to molten steel since manganese can capture sulfur from FeS as shown in Equation (8.4). MnS melts at 1620 ℃, and the hot working temperature of steel is 800 – 1200 ℃. MnS has good plasticity and does not influence the hot working properties of steel.

$$Mn + FeS \Longleftrightarrow MnS + Fe \tag{8.4}$$

6) Other elements

Impurities of N, O, and H are from the air in the production of steel. When molten steel cools down, these elements stay in ferrite as atoms. N increases the strength and hardness of steel while decreases toughness. Al reacts with N to generate AlN. When AlN particles are evenly distributed in the steel, the toughness, plasticity, and abrasion resistance of the steel are increased. When AlN particles are not evenly distributed, the mechanical properties and abrasion resistance of steel are decreased.

After the deoxidation with Mg, Al, and Si, there is still a little oxygen in steel in the form of Al_2O_3, FeO, MnO, and SiO_2, which exist in the steel as chains or strips, especially at the interface of crystals, decreasing the plasticity and toughness of steel.

After hot rolled and forged, atomic hydrogen will gather into a molecular state and appear in the steel. The pressure generated by hydrogen can crack the steel from the inside, forming almost circular flat fracture surfaces, which are called "white spots" or hydrogen blisterings. In particular, some alloy steel, such as ingot steel, shackle steel, shackle inscribed steel, etc., are particularly sensitive to white spots. The hydrogen content can be controlled by using materials of low moisture or hydrogen content, avoiding late addition of lime during the smelting process, minimization of carry-over slag, efficient vacuum degassing, and intense purging.

8.3.2 Metallography

Similar to pure iron, carbon steel has a crystal structure. Crystal phases are defined as regions with the same chemical composition and crystal structure, and have a distinct interface with the surrounding medium. Several different crystalline structures make up the steel at different temperatures.

(1) Ferrite is a solid solution, stable at room temperature, and capable of containing up to only 0.008% carbon. Magnetic ferrite is sometimes called alpha iron.

(2) Cementite is an iron-carbon crystalline (Fe_3C). Cementite contains 6.67% carbon and can combine with ferrite to form pearlite.

(3) Austenite or gamma iron is capable of dissolving 2% carbon. While austenite is never stable in carbon steel at temperatures below 727 °C, additional alloys can make it stable at room temperature. Austenite is nonmagnetic and easily work-hardened.

(4) Pearlite is formed when thin and alternating layers of cementite and ferrite combine. During the slow cooling of an iron-carbon alloy, pearlite is also formed by a eutectoid reaction as austenite cools slowly. Pearlite contains 0.77% carbon, and it usually makes the steel more ductile.

(5) Bainite is hard with low ductility and is a combination of fine carbon needles in a ferrite matrix. It forms when the austenite is cooled at a rate lower than that is needed to form martensite.

(6) Martensite is formed by quenching of the austenite form of iron, in which carbon atoms do not have time to diffuse out of the crystal structure in large enough quantities to form cementite.

8.4 Mechanical Properties

Steel in civil structures is mainly subjected to tension, compression, bending, impaction, and other external loads. The strength, plasticity, bending performance, impact toughness, and hardness are the five important mechanical properties of steel.

8.4.1 Strength

Figure 8.5 illustrates a stress-strain curve of low carbon steel subjected to uniaxial tension. According to specification GB/T 228. 1—2021, during the tensile test, a steel bar is placed in the upper and lower gripping device of the testing machine, and a tensile load is applied until the specimen breaks. The stress-strain curve can be divided into two parts: elastic deformation and plastic deformation.

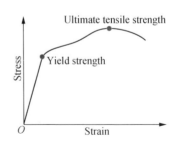

Figure 8.5 Typical uniaxial tensile stress-strain curve for a steel bar

1) Elastic deformation

The first section of the curve, where the stress is a direct proportion of the strain, corresponds to the elastic deformation. The elastic deformation is characterized by total reversibility. The slope of the linear relationship is Young's modulus, E, and is approximately constant for all steels: around 200 GPa.

The area under the stress-strain curve in the elastic region is the energy that the material can absorb reversibly, and is known as resilience. The maximum stress that the material can undergo in the elastic region is the yield strength, σ_s. The yield strength is an important parameter for structural design. In service, the steel must never undergo a stress level higher than the yield strength, otherwise, it will experience permanent deformation.

2) Plastic deformation

Once yield strength has been exceeded, the material enters the plastic deformation region where the relationship between the stress and strain is non-linear. The stress-strain curve then reaches the maximum point which is called the ultimate tensile strength (σ_b) in the plastic deformation region. This ultimate tensile strength represents the start of the rupture process, which leads to material fracture.

Once the ultimate tensile strength has been exceeded, micro cracks begin to appear inside the material and propagate, and deformation stops being uniform. The speed of deformation in the area where the cracks appear is greater than the speed of deformation in other parts of the specimen, due to the concentration of strain around the cracks, thus producing localized deformation. The area of localized deformation in a uniaxial tensile test is the necking and it is where the fracture of the material takes place.

The area under the stress-strain curve is the toughness, which is the energy per unit of volume absorbed by the material from the start of deformation until fracture. Yield strength and ultimate tensile strength are the two important strength characteristics of steel. In structural design, steel members are required to work in an elastic range without plastic deformation. Thus, the yield strength of steel is considered as the design strength of the structure (σ_s). It is expected to use steel with high yield strength and the ratio of yield strength to tensile strength (σ_s/σ_b). Smaller σ_s/σ_b indicates higher reliability and safety factor. Usually, it is suggested to use a σ_s/σ_b value of 0.60 – 0.75. For some metals that do not have a significant yield point, the conditional yield point is typically defined at the 0.2% offset strain. The yield strength at 0.2% offset is determined by finding the intersection of the stress-strain curve with a line parallel to the initial slope of the curve and which intercepts the strain axis at 0.2%.

8.4.2 Plasticity

The ability of steel to undergo a specific permanent deformation before fracture when it is subjected to stress is called plasticity. Plasticity means non-reversible deformation in response to the applied force while ductility is the ability to deform under tensile stress. In Figure 8.6, the left steel bar has low plasticity or ductility and is usually high carbon steel. The right one has better plasticity and is usually alloy steel. The plasticity of steel is of significance in civil engineering. In the manufacturing process, steel with high plasticity can withstand certain forms of external force process without damage. Steel with greater plasticity has better safety. During the service, occasional overload on steel with plasticity can produce plastic deformation to redistribute stress and avoid sudden failure.

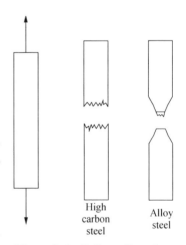

Figure 8.6 Deformation of steel at a tensile fracture

Two values can be calculated based on the tensile strength test to evaluate the plasticity or ductility of steel. As shown in Figure 8.6, elongation of the steel bar over the original length is used to evaluate the ability to withstand permanent deformation at fracture (Equation (8.5)). Reduction of area measures the maximum reduction of the necks in the cross-sectional area, which can be also used to evaluate the relative plasticity or ductility of steel at fracture (Equation (8.6)).

$$\delta_n = \frac{L_1 - L_0}{L_0} \times 100\% \tag{8.5}$$

$$\psi = \frac{A_0 - A_1}{A_0} \times 100\% \tag{8.6}$$

where δ_n = elongation at fracture (%);
L_0 = gauge length (mm);
L_1 = gauge length after fracture (mm);
ψ = reduction of area at fracture (%);
A_0 = original cross-sectional area of the specimen (mm²);
A_1 = cross-sectional area of the specimen after fracture (mm²).

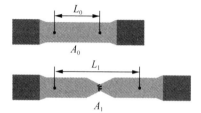

Figure 8.7 Elongation and reduction of area

8.4.3 Cold Bending Property

The cold bending property of steel refers to its ability to withstand bending deformation at normal temperature. Most of the steel bars used in reinforced concrete need to be bent, so they must meet the requirements of cold bending properties. Cold bending tests, like elongation, indicate the plasticity of steel under static loads, but the elongation reflects the plasticity of steel under uniform deformation. Cold bending tests the plasticity or ductility of steel under bending deformation and determines whether there is uneven structure, defects, impurities, etc. It is a simple test measuring the ductility and soundness of steel, which is critical for structural steel.

According to specification GB/T 232—2010 (Figure 8.8), the cold bending property of steel is expressed by the bending angle α and the ratio of the diameter of the bending center d to the diameter (or thickness) of the specimen d_0. The larger the bending angle, the smaller the ratio of the bending diameter to the diameter (or thickness) of the specimen, the better the cold bending property of the steel specimen. After the bending test, the specimen should have no cracks, bedding, or fracture at the bending position.

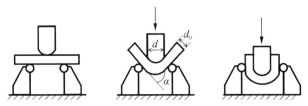

Figure 8.8   Cold bending test

8.4.4   Impact Toughness

The impact toughness of metal is determined by measuring the energy absorbed in the fracture of the specimen. According to specification GB/T 229—2020, it is obtained by noting the height at which the pendulum is released and the height to

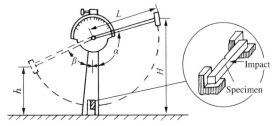

Figure 8.9   Charpy V notch impact test

which the pendulum swings after it has struck the specimen. Figure 8.9 shows the Charpy V-notch impact testing machine and the specimen used for the impact toughness test.

The chemical composition, smelting, and processing quality of steel influence the impact toughness. The high content of phosphorus and sulfur, segregation, non-metallic inclusions, pores and micro-cracks formed in welding, etc., significantly reduce the impact toughness. In addition, the impact toughness of steel is greatly influenced by temperature, and the impact toughness decreases with the decrease of temperature. When the temperature falls below a value, the impact toughness drops sharply, and the steel may exhibit brittle fracture, which is called cold brittleness. Therefore, low-temperature impact toughness must be tested for the steel used in a cold environment, especially the important structure under dynamic loads.

8.4.5   Hardness

Hardness is the ability of steel to resist harder objects pressing into the steel surface. It is usually tested with the Brinell Hardness apparatus. The Brinell Hardness (HB) test applies a predetermined force on a tungsten carbide ball of fixed diameter and holds for a predetermined time, and then removes it. The Brinell Hardness Number (*BHN* or *HB*) is related to the pressure on the indentation area and can be calculated by Equation (8.7).

$$HB = 0.102 \times \frac{2P}{\pi D(D - \sqrt{D^2 - d^2})} \tag{8.7}$$

where *HB* = Brinell Hardness Number;

P = applied specified load (N);

D = ball diameter (mm);

d = indentation diameter (mm).

The diameter of the steel ball is 10 mm, 5 mm, or 2.5 mm. The load and the diameter of the steel ball follow the relationship: $P = 30D^2$. The Brinell Hardness test is accurate and widely used but is not applicable for steel with *HB* higher than 450 and lower thickness. Hardness can also be utilized to estimate the tensile strength of steel as shown in Equation (8.8).

Figure 8.10 Brinell Hardness test

$$\sigma_b = \begin{cases} 0.36HB, & HB \leqslant 175 \\ 0.35HB, & HB > 175 \end{cases} \tag{8.8}$$

8.5 Heat and Cold Treatment

8.5.1 Heat Treatment

Heat treatment is to heat the steel to a specific temperature, hold the temperature for a specified time, and then cool the material at a specified rate (Figure 8.11). It includes four types of treatment.

Figure 8.11 Steel heat treatment

1) Annealing

Annealing is to heat the steel to a specific temperature and cool slowly. There are different types of annealing.

(1) Full annealing is to heat the steel to 50 ℃ above the austenitic temperature line, hold the temperature until all the steel transforms into either austenite or austenite-cementite, and then cool slowly in a furnace. Due to the slow cooling rate, the grain structure is coarse pearlite with ferrite or cementite, and the properties of steel are uniform. The steel is soft and ductile after treatment.

(2) Process annealing is to treat work-hardened parts made with low carbon steel. The material is heated to about 700 ℃ and held long enough to allow recrystallization of the ferrite phase. By keeping the temperature below 727 ℃, there is not a phase shift between ferrite and austenite, and the only change is the refinement of the size, shape, and distribution of the grain structure.

(3) Stress relief annealing is to reduce residual stresses in the cast, welded, cold-worked, and cold-formed parts. The steel is heated to 600 to 650 ℃, held at temperature for about one hour, and then slowly cooled in still air.

(4) Spheroidization is an annealing process used to improve the ability of high carbon steel to be machined or cold worked. It also improves abrasion resistance.

2) Normalizing

Normalizing is similar to annealing with different heating temperatures and cooling rates. It is to heat steel to about 60 ℃ above the austenite line and then cool it in the air for a higher cooling rate. Normalizing produces a uniform, fine-grained microstructure. However, due to the higher cooling rate, the shapes with varying thicknesses result in less uniformity of the normalized parts than annealing. Since structural plate has a uniform thickness, normalizing is an effective process and results in high fracture toughness of the material.

3) Hardening

Hardening is to heat steel to a temperature above the transformation range and then quenched in water or oil for rapid cooling, resulting in high strength and hardness but very low plasticity and toughness. The rapid cooling causes the iron to change into martensite rather than ferrite. Martensite has a very hard and brittle structure. Due to the rapid cooling, the surface of the material is harder and more brittle than the interior of the element, creating non-homogeneous characteristics. Rapid cooling also causes steel pieces with sharp angles or grooves to crack immediately after hardening. Therefore, hardening must be followed by tempering.

4) Tempering

Tempering is after quenching, the steel is cooled to about 40 ℃, then reheated, maintained at the elevated temperature for about two hours, and then cooled in still air to improve ductility and toughness. Heating causes carbon atoms to diffuse from martensite to produce a carbide precipitate and the formation of ferrite and cementite. Tempering is performed to improve ductility and toughness.

8.5.2 Welding

Welding is a technique for joining two metal pieces by applying heat to fuse the pieces. Many civil engineering structures such as steel bridges, frames, and trusses require welding during construction and repair. More than 90% of the connections are welded. There are many types of welding. Generally, arc welding uses an arc between the electrode and the grounded base metal to bring both the base metal and the electrode to their melting points. Gas welding uses an external shielding gas, which shields the molten weld pool and provides the desired arc characteristics.

Due to the high local temperature, steel forms an overheated area in the welding joint. The internal crystal structure changes and tends to produce hard and brittle structures around the welding joint. Materials with low weldability may crack due to the local stresses caused by heating at the weld joint. The weldability of steel is mainly influenced by its chemical composition. When the carbon content exceeds 0.3%, the weldability of steel is poor. The increase of other elements may also reduce the weldability. The welding quality can be improved by preheating before welding and heat treatment after welding, using the correct electrode, and proper operations.

During welding, a small volume of steel melts and cools quickly because of the rapid heat transfer. Complex and non-uniform changes occur, causing expansion, contraction, deformation, internal stress, changes in the crystal structure, and defects. The defects include cracks (thermal cracks), pores, and inclusions such as slag, deoxy products, and nitrous compounds. Defects in the heated zone include cracks (cold cracks) and coarse grains, and precipitate embrittlement. The carbon, nitrogen, and other atoms form carbide or nitrogen compounds, which precipitate and cause increased lattice distortion of the embrittlement. The welding joints require sufficient strength, plasticity, toughness, and fatigue resistance, and the welding quality should be carefully controlled.

8.5.3 Cold Working

Cold working is to draw or roll steel to produce plastic deformation at room temperature, improving strength and surface finishing. After cold-drawn, steel bars are stored at room temperature for 15 – 20 days which is natural aging, or heated to 100 – 200 ℃ and kept for a specific time which is artificial aging. After cold working, the yield strength increases, while the plasticity, toughness, and elastic modulus decrease. As shown in Figure 8.12, the yield strength of the cold-drawn without aging is just a little higher than that of the original steel. However, the cold-drawn steel with aging significantly improves both the yield and ultimate strength of steel after cold-drawn treatment. The proper cold tensile stress and aging treatment measures are usually chosen through tests. Generally, natural aging has more significant effects on the steel bar with lower strength but has less significant effects on the steel bar with higher strength, for which artificial aging must be used.

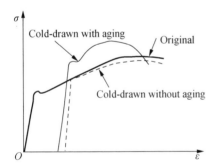

Figure 8.12 Cold-drawn with and without aging

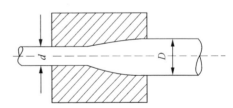

Figure 8.13 Low carbon cold-drawn steel wire

As shown in Figure 8.13, low carbon cold-drawn steel wire is made by letting a round steel bar pass through a smaller area. The cold drawing effect is stronger than the pure drawing action. The cold-drawn wire is not only pulled but also squeezed. Through one or more times of cold drawing, the yield strength can be increased by 40%–60%, but the toughness reduces. According to specification GB 50204—2015, cold-drawn steel wire can be classified into two levels based on strength: Class B for non-prestressed steel wire and Class A for prestressed steel wire.

8.6 Typical Structural Steel

Structural steel includes the steel plate, steel beam, and section steel. Structural Steel used in concrete is generally the steel bar, steel wire, and steel stranded wire for reinforced concrete and prestressed concrete. Specifically, they are the hot-rolled plain steel bars for steel-reinforced concrete (GB/T 1499.1—2017), hot-rolled ribbed steel bars for steel-reinforced concrete (GB/T 20065—2016), steel fibers for prestressed concrete (GB/T 5223—2014), medium-strength steel wires for prestressed concrete (GB/T 30828—2014), steel stranded wires for prestressed concrete (GB/T 5224—2014), cold-rolled ribbed steel bar (GB/T 13788—2017), cold-drawn steel bars and cold-drawn low carbon steel wires. The properties of various steels depend mainly on the type of steel used and the manufacturing method.

8.6.1 Hot-Rolled Steel Bar

The hot-rolled steel bar is one of the most commonly used steel in civil engineering, which can be classified into different types according to its mechanical properties (Table 8.1). There are mainly two types of hot-rolled steel bars: hot-rolled plain bar (HPB) and hot-rolled ribbed bar (HRB). The label is: "HPB" or "HRB" + yield strength. The "E" at the end means it is anti-earthquake.

(1) The HPB is made of Q235 carbon structural steel. It has low strength, but high plasticity and elongation, and is easy for cold working and welding. It is widely used as reinforcing bars in medium- and small-size reinforced concrete structures and tie rods in wood structures. It can be used as the base metal for cold-rolled ribbed steel bars and cold-drawn low carbon steel wires.

(2) The HRB is also used as reinforcing steel (Figure 8.14). It exhibits high strength, good plasticity, and good weldability. On the surface of HRB, the longitudinal ribs and uniformly distributed transverse ribs are ribbed to enhance the bonding between the steel bar and concrete. The ribs include the crescent, contour, herringbone, spiral shapes, etc. Steel can be saved by 40%–50% when using HRB as the load-bearing bars in concrete compared to HPB. Therefore, it is widely utilized as the main steel bar in large- and medium-size reinforced concrete structures.

Table 8.1 Hot-rolled steel bar

Types	Labels	Nominal diameter d_0 (mm)	Min. yield strength (MPa)	Min. tensile strength (MPa)	Min. elongation at fracture (%)	Min. elongation at yield (%)	Diameter d of 180° cold bending test
Hot-rolled plain steel bar	HPB300	6-22	300	420	25	10.0	d_0
Hot-rolled ribbed steel bar	HRB400 HRBF400 HRB400E HRBF400E	6–25	400	540	16	7.5	$4d_0$
		18–40					$5d_0$
		40–50			—	9.0	$6d_0$
	HRB500 HRBF500E	6–25	500	630	15	7.5	$6d_0$
		18–40					$7d_0$
		40–50			—	9.0	$8d_0$
	HRB600	6–25	600	730	14	7.5	$6d_0$
		18–40					$7d_0$
		40–50					$8d_0$

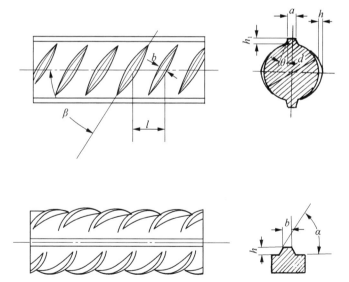

Figure 8.14 The hot-rolled ribbed bar with crescent ribs

8.6.2 Cold-Rolled Ribbed Bar

A cold-rolled ribbed bar (CRB) is formed by reducing the diameter of the hot-rolled bar through cold rolling. Its label is "CRB" + yield strength. "H" at the end means high ductility. The chemical composition and mechanical properties of CRB should satisfy specification GB/T 13788—2017 (Table 8.2). The CRB has high strength, good plasticity, and stable quality. Grade 550 steel bars are mainly used for reinforced concrete structures, especially the main load-bearing bars of slab members and the non-prestressed steel bars in prestressed concrete structures.

Table 8.2 Mechanical properties of CRB

Types	Labels	Min. plastic deformation strength $R_{p0.2}$ (MPa)	Min. tensile strength R_m (MPa)	Min $R_m/R_{p0.2}$	Min. elongation at fracture (%)		Min. elongation at yield (%) A_{gt}	Diameter d of 180° cold bending test	Bending times	Initial stress for stress relaxation Min 1000 h, %
					A	A_{100mm}				
Steel for reinforced concrete	CRB550	500	550	1.05	11.0	—	2.5	$3d_0$	—	—
	CRB600H	540	600	1.05	14.0	—	5.0	$3d_0$	—	—
	CRB680H	600	680	1.05	14.0	—	5.0	$3d_0$	4	5
Steel for prestressed concrete	CRB650	585	650	1.05	—	4.0	2.5	—	3	8
	CRB800	720	800	1.05	—	4.0	2.5	—	3	8
	CRB800H	720	800	1.05	—	7.0	4.0	—	4	5

8.6.3 Hot-Rolled Reinforcing Bar for Prestressed Concrete

The hot-rolled reinforcing bar for prestressed concrete is made by quenching and tempering the hot-rolled low alloy steel bar (Table 8.3). The rebar can be classified into the rounded steel bar, the spiral groove steel bar, the spiral rib steel bar, and the ribbed steel bar. This steel bar is usually rolled into a flexible wire rod with a diameter of greater than 2.0 m. Each wire rod consists of one steel wire with a weight of larger than 700 kg. It features high strength, high toughness, and grip strength, which is mainly employed in prestressed concrete. It should meet the requirements in specification GB/T 5223.3—2017.

The design strength of the hot-rolled reinforcing bar for prestressed concrete is 80% of the standard strength. The tensile stress in the pre- and post-tensioning methods is 70% and 65% of the standard strength, respectively.

Table 8.3 Properties of the hot-rolled reinforcing bar for prestressed concrete

Nominal Diameter (mm)	Grades	Min. yield strength (MPa)	Min. tensile strength (MPa)	Min. elongation at fracture (%)
6	40Si2Mn			
8.2	48Si2Mn	1325	1476	6
10	45Si2Cr			

8.6.4 Cold-Drawn Low Carbon Steel Bar

The cold-drawn low carbon steel bar and wire are widely used cold-drawn steel. The mechanical properties and surface finishes are both improved after cold drawing. After cold drawing under the required elongation and cold bending property, the low carbon steel and low alloy high strength steel have increased yield strength, ultimate strength, and plasticity. However, it should be noted that the steel bar must be cold-drawn after welding; otherwise, the cold-drawn hardening effect will disappear due to the influence of high temperature during welding.

Table 8.4 Mechanical properties of the cold-drawn steel bar

Labels	Cold-drawn			Annealed		
	Min. tensile strength (MPa)	Min. elongation at fracture (%)	Min. area reduction (%)	Min. tensile strength (MPa)	Min. elongation at fracture (%)	Min. area reduction (%)
10	440	8	50	295	26	55
15	470	8	45	345	28	55
20	510	7.5	40	390	21	50
25	540	7	40	410	19	50
30	560	7	35	440	17	45
35	590	6.5	35	470	15	45
40	610	6	35	510	14	40
45	635	6	30	540	13	40

Continued

Labels	Cold-drawn			Annealed		
	Min. tensile strength (MPa)	Min. elongation at fracture (%)	Min. area reduction (%)	Min. tensile strength (MPa)	Min. elongation at fracture (%)	Min. area reduction (%)
50	655	6	30	560	12	40
15Mn	490	7.5	40	390	21	50
50Mn	685	5.5	30	590	10	35
50Mn2	735	5	25	635	9	30

The cold-drawn low carbon steel wire is produced by one or multiple cold-drawn treatments. The mechanical properties of cold-drawn low carbon steel wire are shown in Table 8.5. During the cold-drawn process, the strength of low carbon steel is increased and its elongation rate is decreased because the dislocation in steel increases and prevents the slip of lattice in steel. Therefore, the strength of cold-drawn low carbon steel wire depends on the original strength of hot-rolled steel wire and total deformation after cold-drawn. The time of cold-drawn should be controlled to ensure strength and ductility.

Table 8.5 Mechanical properties of cold-drawn low carbon steel wire

Nominal diameter (mm)	Min. tensile strength (MPa)	Min. elongation at fracture (%)	Bending times	Diameter d of 180° cold bending test (mm)
3	550	2.0	4	7.5
4		2.5		10
5		3.0		15
6				15
7				20
8				20

8.6.5 Prestressed Steel Wire, Indented Steel Bar and Steel Strand

The prestressed steel wire is made of carbon steel of high quality by isothermal quenching and drawing. The diameter of the prestressed steel wire is 2.5 – 5 mm, and the tensile strength is 1500 – 1900 MPa. The surface of the prestressed steel wire can be deformed to enhance the bond with concrete, then into a carved steel wire. The prestressed steel wire has high strength, good plasticity, high corrosion resistance, etc., and is applicable in prestressed concrete.

The indented steel bar and steel strand are all cold-worked reinforced steel, and there is no obvious yield point, so the material inspection can only be based on the tensile strength (Figure 8.15). Its strength is almost twice that of the hot-rolled grade Ⅳ steel bar, and it has good plasticity and can be cut to the required length when used. The design strength is determined by the statistical value of the conditional yield point. The prestressed steel wire, scratched steel wire, and strand steel wire all have the advantages of high strength, high plasticity, and no need for joints. They are applicable to the prestressed concrete structure with a large load, large span, and curved reinforcement.

Figure 8.15　Strand steel wire

8.6.6　Steel Sections

Steel structures in civil engineering usually use different types of steel sections or plates. For example, thin-walled steel structures use thin-walled steel sections, round steel sections, and section angles. Steel sections are well-designed and manufactured steel members to bear specific bending or compressive loads, and can be directly used and assembled in the steel structure quickly and cost-effectively. The steel sections can be connected by bolting, welding, or riveting. Rivets have historically been used as a connecting medium; however, they have largely been replaced by bolts. Usually, steel sections with thick walls (0.35 − 200 mm) are hot-rolled, while steel sections with thin walls (0.2 − 5 mm) are cold-rolled.

1) **Hot-rolled steel sections**

The commonly used section steels of the steel structure include the I-beam, H-beam, T-beam, Z-beam, channel steel, equal-leg angle steel, unequal-leg angle steel, etc. (Figure 8.16). Characteristics of those hot-rolled steel sections include:

(1) I-beam is widely used in various building structures and bridges. It is mainly used for members bearing in the plane of the web, but it should not be used alone as the axial compression member or two-way bending member.

(2) Compared with I-beam, H-beam optimizes the distribution of loads and has a wide flange, large lateral stiffness, strong bending resistance, parallel flange surfaces, and a convenient connection structure. It is often used in large buildings with large bearing capacity and good section stability. H-beam with the wide and middle flange is suitable for axial compression members such as steel columns and H-beam with the narrow flange is suitable for bending members such as steel beams.

(3) Channel steel can be used as members bearing axial force, beams bearing transverse bending, and connecting members. It is mainly used in building structures, vehicle manufacturing, etc.

(4) Angle steel is mainly used as a member and supporting member bearing axial force, and can also be used as a connecting part between stressed members.

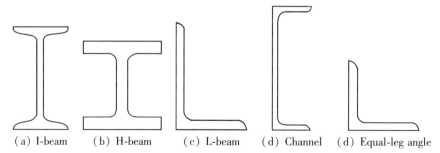

(a) I-beam　　(b) H-beam　　(c) L-beam　　(d) Channel　　(d) Equal-leg angle

Figure 8.16　Typical steel sections

The steel types and grades for the steel structure are mainly selected according to the importance of the structure and member, the type of loads (static or dynamic loads), the connection method (welding or bolting connection), working conditions (ambient temperature and surrounding environment), etc. Carbon structural steel and low alloy steel are mainly used as hot-rolled steel sections. The carbon steel Q235A, low alloy steel Q345 (16Mn), and Q390 (15MnV) are the most widely used. The former is applicable for most steel structures, and the latter can be used for long-span steel structures for dynamic loads. Labels of the hot-rolled steel sections include types, dimensions, steel types, and related specifications.

2) Cold-formed thin-walled steel section

Cold-formed thin-walled steel sections are usually formed by cold bending of 2 – 6 mm thin-walled steel plates. They include angle steel, channel steel and hollow thin-walled sections with square and rectangle shapes, and are mainly used in light steel structures.

3) Steel plate

The steel plates can be classified into hot-rolled and cold-rolled steel plates. According to the thickness, they can be classified into thick steel plates (thickness >4 mm) and thin steel plates (thickness ≤ 4 mm). Thick plates are produced by hot-rolling. Thin plates can be produced by hot-rolling or cold-rolling. Most of the thin plates used in civil engineering are hot rolled.

Steel plates mainly use carbon steel, and some long-span bridges use low alloy steel. Thick steel plates are mainly used for load-bearing members. Thin steel plates are mainly used for roofs, floors, and walls. In steel structures, a single steel plate cannot work independently. Several plates must be connected into I-shaped and box structures to bear the load.

4) Steel pipe

Steel pipes are mostly used to make trusses, tower masts, concrete-filled steel tubes, etc. They are widely used in high-rise buildings, plant columns, tower columns, penstocks, etc. According to the manufacturing process, the steel pipes used for steel structures are classified into hot-rolled seamless steel pipes and welded steel pipes.

(1) The hot-rolled seamless steel pipe is made of high-quality carbon steel and low-alloy structural steel. It is mostly produced by the hot-rolling, cold-drawing, or cold-rolling process. The cost of the latter is high. It is mainly used for penstocks and some specific steel structures.

(2) The welded steel pipe is made of high-quality or ordinary carbon steel plates by cold-welding. According to the welding forms, there are straight seam electric welding steel pipes and spiral welding steel pipes, which are suitable for various structures, transmission pipelines, and other purposes. The welded steel pipe has low cost and is easy to process, but the compressive performance is poor.

8.7 Corrosion and Protection

8.7.1 Corrosion

Steel corrosion is the destruction of its surface by a chemical reaction with the surrounding medium. Corrosion can occur in humid air, soil, and industrial waste gas. The temperature increases, and the corrosion accelerates. Corrosion is critical for all steel structures. When serious corrosion occurs during storage, the cross-sectional area of steel and the quality are reduced. The rust removal work is very costly or can only obtain scrapped steel. If the steel has corrosion in service, the local rust pit can cause stress concentration, accelerating the early damage of the structure. Under repeated loads, corrosion fatigue occurs and the fatigue strength is greatly reduced.

According to different effects of the steel surface and the surrounding environment, corrosion can be classified into two types. Chemical corrosion is the chemical reaction of steel with the surrounding media (such as oxygen, carbon dioxide, sulfur dioxide, and water), generating porous hematite ferrite Ⅲ oxide (Fe_2O_3). Electrochemical corrosion is the contact between the steel and the electrolytic solution that produces an electric current, forming a galvanic cell that causes corrosion. It is the most common type of corrosion. Electrochemical corrosion requires four elements: an anode where corrosion occurs, a cathode to form a corrosion cell, a conductor for electrons to flow, and a liquid electrolyte supporting the flow of electrons (Figure 8.17). Steel contains anodes and cathodes and is also an electrical conductor. Preventing contact with moisture is critical for the prevention of rust.

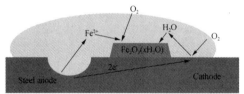

Figure 8.17 Corrosion of steel

8.7.2 Protection

Coating or painting is usually used to prevent the corrosion of steel. Commonly used primers in the paint include red lead, epoxy zinc-rich paint, and iron red epoxy primer. The topcoats include grey lead oil and phenolic enamel. For thin steel products, hot-dip galvanized coating is very effective while costly.

Concrete is strong alkalinity, with a pH value of around 12.5, forming alkaline protection on steel. There are several reasons for the corrosion of steel bars in concrete. Concrete contains voids and micro cracks, and the thickness of the protective concrete layer might be thin or heavily carbonized. In particular, when chloride iron content is too high, the corrosion of steel is severe. To improve the corrosion resistance of reinforced concrete, barrier coating or sacrificial primers are applied. Besides, the water-cement ratio and cement content should be controlled. The concrete should also be thick enough to prevent steel corrosion. It is also important to control the chloride content in additives for concrete.

Prestressed steel bar is susceptible to corrosion as it has a high content of carbon and is subjected to deformation and cold-drawn process. In particular, the prestressed steel after high strength hot-treated tends to produce corrosion. Thus, for a prestressed loading structure, the quality of raw materials should be strictly controlled and chloride should be avoided.

To prevent corrosion of the reinforced steel bar, it is suggested to use a rust inhibitor. Moreover, galvanization, cadmium, and nickel coating can also effectively enhance the corrosion resistance of the reinforced steel bar.

Questions

1. What is the effect of carbon and other elements on the mechanical properties of steel?
2. What is the typical maximum percentage of carbon in the steel?
3. Define yield strength and ultimate tensile strength of steel.
4. What is the hardness of steel and can we use hardness instead of strength?
5. Why is impact toughness important for steel? What's the effect of temperature on the impact toughness?
6. How to measure the plasticity of steel?
7. Define the alloy steel. Explain why alloys are added to steel.
8. Discuss the procedures and objectives of the four heat treatments.
9. What is the effect of cold-drawn treatment?
10. How to prevent the corrosion of steel?

Reference

[1] Doran D, Cather B. Construction materials reference book[M]. Abingdon: Routledge, 2013.

[2] Illston J M, Domon P. Construction materials: their nature and behaviour[M]. 3rd ed. Boca Raton, FL: CRC Press, 2001.

[3] Mamlouk M S, Zaniewski J P. Materials for civil and construction engineers[M]. 4th ed. Hoboken, NJ: Pearson, 2017.

[4] Gonçalves M C, Margarido F. Materials for construction and civil engineering[M]. Cham: Springer International Publishing, 2015.

[5] Claisse P A. Civil engineering materials[M]. Oxford: Butterworth-Heinemann, 2016.

[6] Brown E R, Kandhal P S, Roberts F L, et al. Hot mix asphalt materials, mixture design, and construction[M]. 3rd ed. Lanham, MD: NAPA Research and Education Foundation, 2009.

[7] Papagiannakis A T, Masad E A. Pavement design and materials[M]. Hoboken, NJ: John Wiley & Sons, Inc. 2008.

[8] Kosmatka S H, Kerkhoff B, Panarese W C. Design and control of concrete mixtures[M]. 14th ed. Skokie, IL: Portland Cement Association, 2002.

[9] Thom N. Principles of pavement engineering[M]. London: Thomas Telford, 2008.

[10] Huang X M, Gao Y, Zhou Y. Civil engineering materials[M]. Nanjing: Southeast University Press, 2020.

[11] Shen A Q. Road engineering materials[M]. Beijing: China Communications Press, 2016.

[12] Li L H, Zhang N W. Road constructional materials[M]. Beijing: China Communications Press, 2006.

[13] Tab Y Q. Asphalt and asphalt mixture[M]. Harbin: Harbin Institute of Technology Press, 2007.

[14] Huang W R, XIONG C H. Asphalt and asphalt mixture[M]. Beijing: China Communications Press, 2020.

[15] Shen A Q, Guo Y C. Cement and concrete[M]. Beijing: China Communications Press, 2019.

[16] Hu X P, Liu J X. Construction materials[M]. Beijing: Peking University Press, 2018.

[17] Wang S F. Construction materials[M]. Wuhan: Wuhan University Press, 2000.

[18] Zhao P, Xie F Z, Sun W S. Fundamentals of materials science[M]. Harbin: Harbin Institute of Technology Press, 1999.

[19] China Architecture & Building Press. General specification of current building materials[M]. Beijing: China Architecture & Building Press, 2000.

[20] AASHTO. Standard method of test for resistance to degradation of small-size coarse aggregate by abrasion and impact in the Los Angeles Machine (AASHTO T 96—02)[S]. Washington, D.C.: American Association of State Highway and Transportation Officials, 2002.

[21] AASHTO. Standard method of test for determining the resilient modulus of soils and aggregate materials (AASHTO T 307—99)[S]. Washington, D.C.: American Association of State Highway and Transportation Officials, 1999.

[22] AASHTO. Standard specification for viscosity-graded asphalt cement (AASHTO M 226—80)[S]. Washington, D.C.: American Association of State Highway and Transportation Officials, 1980.

[23] AASHTO. Standard specification for performance-graded asphalt binder (AASHTO M 320—10)[S]. Washington, D.C.: American Association of State Highway and Transportation Officials, 2010.

[24] AASHTO. Standard specifications for transportation materials and methods of sampling and testing (AASHTO T324—19)[S]. Washington, D.C.: American Association of State Highway and Transportation Officials, 2019.

[25] AASHTO. Method of test for determining rutting susceptibility of hot mix asphalt (HMA) using the asphalt pavement analyzer (AASHTO T 340—10)[S]. Washington, D.C.: American Association of State Highway and Transportation Officials, 2010.

[26] AASHTO. Standard method of test for determining the creep compliance and strength of hot mix asphalt (HMA) using the indirect tensile test device (AASHTO T 322—07)[S]. Washington, D. C.: American Association of State Highway and Transportation Officials, 2007.

[27] AASHTO. Standard method of test for determining the dynamic modulus and flow number for asphalt mixtures using the asphalt mixture performance tester (AMPT) (AASHTO T 378—17)[S]. Washington, D. C.: American Association of State Highway and Transportation Officials, 2017.

[28] AASHTO. Standard method of test for resistance of compacted asphalt mixtures to moisture-induced damage (AASHTO T 283—07)[S]. Washington, D. C.: American Association of State Highway and Transportation Officials, 2007.

[29] AASHTO. Standard practice for superpave volumetric design for asphalt mixtures (AASHTO R 35—17)[S]. Washington, D. C.: American Association of State Highway and Transportation Officials, 2017.

[30] ASTM. Standard test method for resistance to degradation of small-size coarse aggregate by abrasion and impact in the Los Angeles machine (ASTM C131/C131M—20)[S]. West Conshohocken, PA: American Society for Testing and Materials International, 2020.

[31] ASTM. Standard terminology relating to concrete and concrete aggregates (ASTM C125—21)[S]. West Conshohocken, PA: American Society for Testing and Materials International, 2021.

[32] ASTM. Standard test method for flat particles, elongated particles, or flat and elongated particles in coarse aggregate (ASTM D4791—19)[S]. West Conshohocken, PA: American Society for Testing and Materials International, 2019.

[33] ASTM. Standard test method for compressive strength of cylindrical concrete specimens (ASTM C39/C39M—21)[S]. West Conshohocken, PA: American Society for Testing and Materials International, 2021.

[34] ASTM. Standard test method for splitting tensile strength of cylindrical concrete specimens (ASTM C496/C496M—11)[S]. West Conshohocken, PA: American Society for Testing and Materials International, 2011.

[35] ASTM. Standard specification for coal fly ash and raw or calcined natural pozzolan for use in concrete (ASTM C618—19)[S]. West Conshohocken, PA: American Society for Testing and Materials International, 2019.

[36] ASTM. Standard specification for viscosity-graded asphalt cement for use in

pavement construction (ASTM D3381/3381M—18)[S]. Washington, D. C.: American Society for Testing and Materials International, 2018.

[37] Ministry of Transport of the People's Republic of China. Test Methods of Aggregate for Highway Engineering (JTG E42—2005)[S]. Beijing: China Communications Press, 2005.

[38] Ministry of Transport of the People's Republic of China. Technical Guidelines for Construction of Highway Roadbases (JTG/T F20—2015)[S]. Beijing: China Communications Press, 2015.

[39] Ministry of Transport of the People's Republic of China. Test Methods of Cement and Cement Concrete for Highway Engineering (JTG 3420—2020)[S]. Beijing: China Communications Press, 2020.

[40] Ministry of Transport of the People's Republic of China. Test Methods of Materials Stabilized with Inorganic Binders for Highway Engineering (JTG E51—2009)[S]. Beijing: China Communications Press, 2009.

[41] Ministry of Transport of the People's Republic of China. Standard Test Methods of Bitumen and Bituminous Mixtures for Highway Engineering (JTG E20—2011)[S]. Beijing: China Communications Press, 2011.

[42] Ministry of Transport of the People's Republic of China. Technical Specification for Construction of Highway Asphalt Pavements (JTG F40—2004)[S]. Beijing: China Communications Press, 2004.

[43] Ministry of Industry and Information Technology of the People's Republic of China. Building quicklime (JC/T 479—2013)[S]. Beijing: China Architecture & Building Press, 2013.

[44] Ministry of Industry and Information Technology of the People's Republic of China. Building hydrated lime (JC/T 481—2013)[S]. Beijing: China Architecture & Building Press, 2013.

[45] General Administration of Quality Supervision, Inspection and Quarantine of the People's Republic of China, Standardization Administration of the People's Republic of China. Common Portland Cement (GB 175—2007)[S]. Beijing: Standards Press of China, 2007.

[46] General Administration of Quality Supervision, Inspection and Quarantine of the People's Republic of China, Standardization Administration of the People's Republic of China. Test methods for water requirement of normal consistency, setting time and soundness of the Portland cement (GB/T 1346—2011)[S]. Beijing: Standards Press

of China, 2011.

[47] General Administration of Quality Supervision, Inspection and Quarantine of the People's Republic of China, Standardization Administration of the People's Republic of China. Fly ash used for cement and concrete (GB/T 1596—2017)[S]. Beijing: Standards Press of China, 2017.

[48] General Administration of Quality Supervision, Inspection and Quarantine of the People's Republic of China, Standardization Administration of the People's Republic of China. Test method for analysis of coal ash (GB/T 1574—2007)[S]. Beijing: Standards Press of China, 2007.

[49] General Administration of Quality Supervision, Inspection and Quarantine of the People's Republic of China, Standardization Administration of the People's Republic of China. Quality carbon structure steels (GB/T 699—2015)[S]. Beijing: Standards Press of China, 2015.

[50] General Administration of Quality Supervision, Inspection and Quarantine of the People's Republic of China, Standardization Administration of the People's Republic of China. Alloy structure steels (GB/T 3077—2015)[S]. Beijing: Standards Press of China, 2015.

[51] General Administration of Quality Supervision, Inspection and Quarantine of the People's Republic of China, Standardization Administration of the People's Republic of China. Structural steel for bridge (GB/T 714—2015)[S]. Beijing: Standards Press of China, 2015.

[52] General Administration of Quality Supervision, Inspection and Quarantine of the People's Republic of China, Standardization Administration of the People's Republic of China. Metallic materials—Tensile testing—Part 1: Method of test at room temperature (GB/T 228.1—2010)[S]. Beijing: Standards Press of China, 2010.

[53] General Administration of Quality Supervision, Inspection and Quarantine of the People's Republic of China, Standardization Administration of the People's Republic of China. Metallic materials - Bend test (GB/T 232—2010)[S]. Beijing: Standards Press of China, 2010.

[54] General Administration of Quality Supervision, Inspection and Quarantine of the People's Republic of China, Standardization Administration of the People's Republic of China. Steel for the reinforcement of concrete—Part 1: Hot rolled plain bars (GB/T 1499.1—2017)[S]. Beijing: Standards Press of China, 2017.

[55] General Administration of Quality Supervision, Inspection and Quarantine of

the People's Republic of China, Standardization Administration of the People's Republic of China. Screw-thread steel bars for the prestressing of concrete (GB/T 20065—2016)[S]. Beijing: Standards Press of China, 2016.

[56] General Administration of Quality Supervision, Inspection and Quarantine of the People's Republic of China, Standardization Administration of the People's Republic of China. Steel wire for prestressing of concrete (GB/T 5223—2014)[S]. Beijing: Standards Press of China, 2014.

[57] General Administration of Quality Supervision, Inspection and Quarantine of the People's Republic of China, Standardization Administration of the People's Republic of China. Middle strength steel wire for prestressed concrete (GB/T 30828—2014)[S]. Beijing: Standards Press of China, 2014.

[58] General Administration of Quality Supervision, Inspection and Quarantine of the People's Republic of China, Standardization Administration of the People's Republic of China. Steel strand for prestressed concrete (GB/T 5224—2014)[S]. Beijing: Standards Press of China, 2014.

[59] General Administration of Quality Supervision, Inspection and Quarantine of the People's Republic of China, Standardization Administration of the People's Republic of China. Cold rolled ribbed steel bars (GB/T 13788—2017)[S]. Beijing: Standards Press of China, 2017.

[60] General Administration of Quality Supervision, Inspection and Quarantine of the People's Republic of China, Standardization Administration of China. Steel bars for the prestressing of concrete (GB/T 5223.3—2017)[S]. Beijing: Standards Press of China, 2017.

[61] General Administration of Quality Supervision, Inspection and Quarantine of China, Standardization Administration of the People's Republic of China. Metallic costings—Electrodeposited coatings of nickel (GB/T 9798—2005)[S]. Beijing: Standards Press of China, 2005.

[62] Ministry of Housing and Urban-Rural Development of the People's Republic of China. Standard for test method of performance on ordinary fresh concrete (GB/T 50080—2016)[S]. Beijing: China Architecture & Building Press, 2016.

[63] Ministry of Housing and Urban-Rural Development of the People's Republic of China. Specification for mix proportion design of ordinary concrete (JGJ 55—2011)[S]. Beijing: China Architecture & Building Press, 2011.

[64] Ministry of Housing and Urban-Rural Development of the People's Republic

of China, State Administration for Market Regulation. Standard for test methods of concrete physical and mechanical properties (GB/T 50081—2019) [S]. Beijing: China Architecture & Building Press, 2019.

[65] Ministry of Housing and Urban-Rural Development of the People's Republic of China. Standard for evaluation of concrete compressive strength (GB/T 50107—2010) [S]. Beijing: China Architecture & Building Press, 2010.

[66] Ministry of Housing and Urban-Rural Development of the People's Republic of China. Standard for test methods of long-term performance and durability of ordinary concrete (GB/T 50082—2009) [S]. Beijing: China Architecture & Building Press, 2009.

[67] Ministry of Housing and Urban-Rural Development of the People's Republic of China. Standard for design of concrete structure durability (GB/T 50476—2019) [S]. Beijing: China Architecture & Building Press, 2019.

[68] Ministry of Housing and Urban-Rural Development of the People's Republic of China. Code for acceptance of constructional quality of concrete structures (GB 50204—2015) [S]. Beijing: China Architecture & Building Press, 2015.

[69] General Administration of Quality Supervision, Inspection and Quarantine of the People's Republic of China, Standardization Administration of the People's Republic of China. Method of testing cements—Determination of strength (GB/T 17671—1999) [S]. Beijing: Standards Press of China, 1999.

[70] State Administration for Market Regulation, Standardization Administration of China. Metallic materials—Charpy pendulum impact test method (GB/T 229—2020) [S]. Beijing: Standards Press of China, 2020.

[71] General Administration of Quality Supervision, Inspection and Quarantine of the People's Republic of China. Carbon structural steels (GB/T 700—2006) [S]. Beijing: Standards Press of China, 2006.

[72] State Administration for Market Regulation, Standardization Administration of China. High strength low alloy structural steels (GB/T 1591—2018) [S]. Beijing: China Quality Inspection Press, 2018.

[73] Ministry of Housing and Urban-Rural Development of the People's Republic of China. Specification for mix proportion design of masonry mortar (JGJ/T 98—2010) [S]. Beijing: China Architecture & Building Press, 2010.